STO

338.
YOSHIKAWA,
SCIENCE HAS NO NATIONAL
BORDERS

ALLEN COUNTY PUBLIC LIBRARY
FORT WAYNE, INDIANA 46802

You may return this book to any location of
the Allen County Public Library.

DEMCO

Science Has No National Borders

Science Has No National Borders

Harry C. Kelly and the Reconstruction of
Science and Technology in Postwar Japan

Hideo Yoshikawa and Joanne Kauffman

The MIT Press
Cambridge, Massachusetts
London, England

© 1994 Massachusetts Institute of Technology and Mita Press.

All rights reserved. No part of this book may be reproduced in any form or by any electronic or mechanical means (including photocopying, recording, or information storage and retrieval) without permission in writing from the publisher.

Set in Bembo by DEKR Corporation.
Printed and bound in the United States of America.

Library of Congress Cataloging-in-Publication Data

Yoshikawa, Hideo.
 Science has no national borders : Harry C. Kelly and the reconstruction of science and technology in postwar Japan / Hideo Yoshikawa and Joanne Kauffman.
   p. cm.
 Includes bibliographical references and index.
 ISBN 0-262-24037-8
 1. Science—Japan—History. 2. Technology—Japan—History. 3. Science and state—Japan—History. 4. Technology and state—Japan—History. 5. Kelly, Harry C.—Influence. 6. Japan—History—Allied occupation, 1945–1952.
 7. Reconstruction (1939–1951)—Japan. 8. Physicists—United States—Biography. I. Kauffman, Joanne. II. Title.
 Q127.J3Y677 1994
 338.95206—dc20   93-29741
                                                                        CIP

Contents

Preface to the American Edition
Joanne Kauffman — vii

Foreword to the Japanese Edition
Takashi Mukaibo — xi

Preface to the Japanese Edition
Hideo Yoshikawa — xiv

1
Japanese Science, 1945 — 1

2
Forces for Change in Japan's Science Policy — 19

3
A U.S. Model for Japan? — 43

4
A Japanese Framework for Science Policy — 61

5
Kelly and Nishina — 79

6
Bridging Two Worlds   95

Epilogue   105

Appendix: Remarks to the Science Council of Japan
on the 25th Anniversary
Harry C. Kelly   107

Notes   117

Index   135

Preface to the American Edition

Joanne Kauffman

From 1945 through 1952 the United States occupied Japan. The mission of the occupation was not to subjugate the country but to help it transform and revitalize its institutions. The difficulties inherent in forging an alliance with a former enemy were compounded by the fact that many of the Americans involved were young and were inexperienced in Japan's language and culture. Nevertheless, some of the Americans who served in the occupation were able to create strong personal ties that were to prove crucial to the success of the mission. Nowhere was this more true than in the reconstruction of Japanese science and the reintegration of Japanese scientists into the global community of scholars after the war.

The story of Harry C. Kelly and the relationships he built with Japanese scientists through his work in the Scientific and Technical Division sheds interesting light on this aspect of the occupation. It is a curious fact that Kelly's name—still greatly respected in the Japanese scientific community—is virtually unknown in the United States. Moreover, despite our high level of curiosity concerning Japan's scientific and technological achievements of the last 20 years, the Scientific and Technical Division, which played a key role in rehabilitating Japan's capacity for scientific inquiry, has been lost to our collective memory.

The paucity of knowledge about Kelly's division is particularly unfortunate in the present climate. Today, the strength of U.S.-Japan relations is being challenged more than at any time in the past 50 years. While officially the ties that were forged during the occupation remain strong, there can be little doubt that they have been stretched taut. As both countries seek to redefine their roles in the post-Cold War order, voices of cooperation and trust are increasingly being drowned out by choruses of suspicion and national self-interest. In redirecting attention to the work of individuals who faced equal if not greater obstacles to mutual trust, this book suggests that an alternative to the present state of affairs is possible.

Harry Kelly was a physicist who believed strongly in the basic scientific principle of openness. The main question that he and his colleagues in the Scientific and Technical Division faced was this: What is the appropriate role of science in the reconstruction of Japan's society and in the rebuilding of its economy for peaceful purposes? Their answer was that science would be pivotal, but that it could contribute only if the old hierarchical institutional structures were replaced.

Today Japan is being asked by many, both at home and abroad, to rethink its policies for science and technology once again. As was the case at mid-century, the challenge goes to the very core of Japan's traditions. The recent calls for more openness in research and development seem, in fact, reminiscent of the demand issued from Commodore Perry's "Black Ships," anchored in the Bay of Tokyo in the summer of 1863, that Japan open its markets to trade with the West. For the next 150 years, Japan sought quite openly to catch up with and surpass the West. Now that it *has* caught up, both Japan and the United States face new challenges in adapting to Japan's role as an international power. The joint publication of this book by Mita Press and The MIT Press reminds us of the role that scientists can play in fostering the cooperation that is needed to meet those challenges.

Written for a Japanese audience, the original edition of this book required some restructuring as well as the addition of information for English-language readers, who are likely to be unfamiliar with the functions of the occupation bureaucracy in general and with those of the Scientific and Technical Division in particular. To my surprise, as I began to research the English-language literature on the occupation, I found few references to science and technology. Fortunately, I found a rich and largely untapped collection of documents in the extensive archives of the occupation in Suitland, Maryland, and in the archives of the National Academy of Sciences in Washington I found a great deal of interesting material on the postwar relations between American and Japanese scientists. I drew extensively on those sources when studying the NAS's role in advising Japanese scientists on their reorganization efforts, and I want to thank NAS archivist Janice Goldblum for her valuable guidance and assistance in this phase of my research.

For information about Harry Kelly's personal background and on his professional life after the occupation, I am particularly grateful to his son Henry and his family, who welcomed me into their home and shared stories, letters, and photographs. Maurice Toler, Director of the North Carolina State University Archives, was most helpful in making available Harry Kelly's papers on file there.

I have relied heavily on the research of John Dower and Charles Weiner for background information on the use and the destruction of the Japanese cyclotrons and on the significance of that destruction for the work of the Scientific and Technical Division. Biographical references and Harry Kelly's personal observations about the occupation are taken largely from an oral history interview of Kelly by Charles Weiner. The interview was made available to me by the archives of the Massachusetts Institute of Technology. Samuel Coleman's study of Kelly and the Physical and Chemical Research Institute underpins this entire endeavor. I am especially grateful to Richard Samuels and Suzanne Berger, who introduced me to the project and provided invaluable guidance. Justin Bloom, former Science Attaché to the U.S. Embassy in Japan, graciously read and commented on an earlier version of the manuscript and suggested fruitful research sources. Bowen Dees, who served with Harry Kelly in the occupation and later at the National Science Foundation, patiently and generously shared his experiences with me in lengthy telephone calls. I thank all my readers and advisers for their help in pointing out errors.

This book was made possible by a generous grant provided by Mita Press. Grateful acknowledgment is due to those at Mita Press and at The MIT Press who have encouraged and aided this work. Deep appreciation is also due Masao Yoshida for his skillful translation of the original book. In carrying out my research, I have been generously supported by MIT's Department of Political Science.

Foreword to the Japanese Edition

Takashi Mukaibo
President Emeritus, University of Tokyo

My memories of Harry Kelly go back to the summer of 1954, when I arrived in Washington to take up my assignment as the first Science Attaché appointed by Japan. I was met at the airport by Dr. Kelly, then Assistant Director of the National Science Foundation, and his colleague Dr. Bowen Dees. Earlier, when Dr. Kelly was in Japan with the Scientific and Technical Division of the Economic and Scientific Section of the Allied Forces, I was too young to have met him; I learned from others what Japanese scientists owed him.

The first thing a science attaché must learn upon taking office is what institutions and persons to contact to ask specific questions in any domain of science. In this regard, Dr. Kelly's boundless kindness was a blessing without which I could never have so smoothly entered into my functions. Dr. Kelly first explained to me the role and activities of the NSF. I was then introduced to the president of the National Academy of Sciences, Dr. Detlev W. Bronk, from whom I received a standing invitation to attend, as informal guest member, the NAS's monthly lecture meetings and social gatherings. This permitted me quickly to gain acquaintances and friends among eminent U.S. scientists. I owed to Dr. Kelly's generous offices almost all my early contacts in academic and administrative circles.

At that time, not many countries sent science attachés to Washington; they numbered seven or so. Through these colleagues I was also able to learn more about the functions and duties of my position.

My family had had no experience living abroad, so I took the advice of the Foreign Office to leave them in Japan while I got established in Washington. When my family joined me, three months later, it was Mrs. Kelly's turn to take us under her wings, to help accustom us to living abroad. It happened that the Kelly family's second son, Tommy, was the same age as my eldest boy. They had two sons, and we had three. The five surprised their families by ignoring the language barrier and becoming friends long before our children started to speak English.

What seemed like extraordinary kindnesses to me and my family were unexceptional for Dr. Kelly. My family and I were far from being exceptions to his usual consideration: countless Japanese scientists benefited from his care over the years.

In the 1960s, the Japan–United States Committee on Scientific Cooperation, based on an agreement between Prime Minister Ikeda and President Kennedy, held its first meetings, co-chaired by Dr. Kelly and Dr. Kankuro Kaneshige. Regular meetings of the committee to determine the cooperative research agenda continue to this day. The NSF and the Japan Society for the Promotion of Science administer the grants on the two sides. What is being achieved by developing joint research far exceeds the monetary investment in the two countries. Today, more than 200 scientific papers are published every year through the activities of the Committee.

The United States has concluded scientific-cooperation agreements with many countries, but according to the National Science Foundation the agreement with Japan is considered to be the most successful and has served as a model for similar arrangements with other nations. This success is, of course, due to the efforts of many scientists and to the devoted service of the secretariats in both countries; however, the greatest factor in the committee's success is undoubtedly the assiduous guiding spirit of its first co-chairmen, Drs. Kelly and Kaneshige. When the two retired from their co-chairmanship, they were presented with awards in recognition of their services. The United States presented Dr. Kaneshige with an honorary diploma, and the Japanese government decorated Dr. Kelly with the Second Order of the Sacred Treasure.

Dr. Kelly considered U.S. cooperation with Japan his highest aim, and he never forgot Japan throughout his career. In 1985, a suggestion was made in Japan to publish a book recording his work while individuals who remembered Dr. Kelly personally were still alive. About the same time, in the United States, a proposal was put forward to start a commemorative undertaking. It was finally agreed that we in Japan would produce a videotape of recollections of Dr. Kelly's life and publish this book. In the United States, a Harry C. Kelly Memorial Fund would be created, with 50 percent of the funding from Japan, to support the Japan Center Program in Science and Technology on the campus of North Carolina State University, where Dr. Kelly had served as dean of students and then provost from 1963 until his retirement in 1974.

It is my sincere wish that this book will inform many people about Dr. Kelly and what Japan owes to him. I sincerely hope, too, that some day Japan, now that it has come to play such a prominent role in the world community, will in turn produce young leaders who, like Dr. Kelly, will think globally of science and its impact on the world community, and effectively contribute to its promotion.

Preface to the Japanese Edition

Hideo Yoshikawa

When I joined the staff of the Japan Atomic Energy Research Institute, Drs. Sagane and Nishibori were among its directors, and Masao Yoshida was my direct superior. They often spoke about Harry Kelly. Later, when I became interested in the history of the development of nuclear energy in Japan, the role played by Dr. Kelly in the rehabilitation of postwar Japanese science, although not directly related to nuclear work, renewed my interest in him.

My thoughts were shared by Genya Chiba of the Research Development Corporation of Japan, who was then exploring ways to reform the conduct of scientific research in Japan. His aim was to incorporate into the present system, which focuses on organizations, a flexible concept of research focused on individual scientists. The ultimate target was to stimulate and promote creative basic scientific research.

What was behind Chiba's thinking was that the era of importing ready-made technology was over and that Japan must now nurture its own innovative technologies. For this, a new organization for scientific research was needed.

Chiba came across Kelly's name while doing research on the postwar rehabilitation of Japanese scientific research. Some time after this, I joined Chiba to assist him in launching the promotion of creative science and technology, and by chance we discovered our mutual interest in the history of Dr. Kelly.

In April 1985 we approached Professor Mukaibo with the idea of writing a book about Kelly for the benefit of young scientists and engineers. Mukaibo agreed to the idea, but he warned us that in an age when even many university students were more accustomed to comic books such a volume could not be expected to sell very well. But then we met a promising ally, Yoshihiro Mita, president of Mita Press. "If it's such a worthy book, I will publish it," he said. Mita was moved by Chiba's enthusiasm for the reform of Japanese science and technology. After several consultations between the two it was agreed that, since I had already assembled some appropriate reference material, I would compile the book. At this point we met another important ally, Dr. Samuel Coleman, then Associate Director of the Japan Center at North Carolina State University, where Kelly had served as provost. Coleman had not known Kelly personally, but through his research on U.S.-Japan relations in science and technology he had come to regard Kelly as a key figure. He was also organizing a fund in Kelly's name to promote bilateral exchanges of young scientists. In June 1985, Coleman visited Japan to present his project to prospective participants. His visit provided the opportunity for U.S. and Japanese groups to get together on the matter of publishing Kelly's biography.

As discussions proceeded, it was suggested that a television documentary on Kelly's activities in Japan might be the best way to reach the general public. Production of the piece was entrusted to the Televiman Union. Location trips to the United States and to Hokkaido were made, and the hour-long film, entitled "Dr. Kelly's Legacy," was broadcast over the Asahi Network on February 11, 1986.

Two weeks before the broadcast, on January 31, Mita Press arranged for a preview and a memorial gathering of Dr. Kelly's friends and his son Henry. A number of people who had known

Kelly 40 years earlier participated in the event and were unanimous in their desire to keep his memory alive. This book is the fruit of their encouragement and help.

My role in preparing the book was not to assemble, arrange, and compile, but simply to write up what was readily and willingly made available to me by those who sincerely wished to see such a book published. Grateful acknowledgment is due in particular to Dr. Takeshi Mukaibo for his contribution of the foreword, to Dr. Sam Coleman for making available documentation from archives in the United States, to Mr. G. Chiba for guiding the entire project, from the television broadcast and the memorial gathering to the publication of the present book, and to Mr. Y. Mita of Mita Press for sponsoring the project.

In the United States, the National Science Foundation and North Carolina State University most courteously provided well-organized documents and photographs for translation and reproduction. I am also deeply indebted to the many institutions and establishments that accorded their kind permission to cite or make reference to reports that include information on Dr. Kelly, and to the institutions with which he was associated in his professional life.

Science Has No National Borders

Japanese Science, 1945

1

Just before he flew from Okinawa to Tokyo on August 30, 1945, to take command of the Allied Forces Occupation of Japan, General Douglas MacArthur, Supreme Commander of the Allied Powers (SCAP), received his orders from Washington. The Post-Surrender Policy Directive prepared by the State-War-Navy Coordinating Committee (SWNCC 150/4) provided few details but made it clear that the United States intended to change Japan.

The two basic goals of the occupation were to demilitarize and to democratize the conquered nation. The directive called for significant involvement in institution building in virtually every social, political, and economic aspect of society. It was followed up on September 22 with more precise military orders that provided a framework and a rationale for all the occupation planning that followed. Setting a stern but constructive tone that emphasized demilitarization, the orders clearly indicated that the intent of the United States was not to destroy Japan but to reform it.[1]

In the early weeks of September, after the formal surrender ceremonies, U.S. forces began arriving in Tokyo to set up the occupation bureaucracy that would remain in place until April 1952. At first there was talk of direct military government by the U.S. Army, as in Germany; however, in the face of daunting cultural and linguistic challenges, this was ruled out in favor of indirect administration through the existing Japanese government.

Indirect involvement meant that the American civilians who were working with the Japanese could influence the pace and the direction of change. Between 1946 and 1948 the number of American officers and civilians on MacArthur's staff grew from around 1,500 to about 3,200. Of these, according to one estimate, only about 500 were in managerial and decision-making positions.[2] Within a month of MacArthur's arrival, the General Headquarters (GHQ) for the occupation forces was organized into four military staff sections and eleven administrative sections. The Economic and Scientific Section (ESS) was the largest. The others were Public Health and Welfare, Civil Information and Education, Civil Intelligence, Statistical and Reports, Civil Intelligence, Legal, Civil Communications, Natural Resources, Government, and Civilian Personnel. At its peak, the occupation's work force (including Japanese civilians serving as translators and as clerical employees) would number more than 300,000.

Much has been written about the Allied occupation of Japan— usually from the perspective of high-level decision makers and in relation to major historical events.[3] By contrast, little has been written about the impact mid-level U.S. personnel may have had through their personal contacts and working relationships with the Japanese.[4] Many of the most important stories thus remain untold, and this is one reason why the occupation remains a curiously enigmatic period in the history of both Japan and the United States.

In this book we shall look at the occupation through the involvement of Harry C. Kelly, a young American physicist who was part of a small but important office in the vast occupation bureaucracy: the Scientific and Technical Division of the Economic and Scientific Section. This office was important to the Americans because they needed scientific expertise to evaluate the military potential of Japanese science and technology, and it was important to the Japanese because they were determined to learn everything they could from their occupiers. Whereas the goals of the United States were to demilitarize and democratize Japan, the Japanese were still guided, as they had been since the nineteenth century, by the desire to catch up with and even surpass the West. But in the autumn of 1945 the most pressing issues facing Japan's leaders had to do with survival and economic recovery. As bureaucratic and industrial elites jockeyed for influence with decision-makers in the occupation bureaucracy, many of Japan's scientists pondered the future of their country's scientific establishment. For some, the realization that massive institutional reform was underway presented an opportunity to rejuvenate a community that they believed had become tradition-bound and unable to promote the advancement of modern science. To them, the occupation's goal of democratizing Japanese institutions mirrored their own aspirations for the future of Japanese science.

The War and U.S. Science Policy

Two massive wartime projects, the Radiation Laboratory at the Massachusetts Institute of Technology (where Harry Kelly had worked) and the Manhattan Project, had clearly demonstrated how an alliance between government and science could advance national interests. The peacetime implications of such an alliance were a source of intense discussion and debate in both the United States and Japan. Many scientists who had been caught up in the war effort were eager to return to their scientific pursuits and reopen free discussions with colleagues around the world. But how should science be organized now that peace had been attained at such a high cost? What was the proper relation between science and government in the atomic era? What were the rights and responsibilities of scientists? Scientists and policy makers in both countries now found themselves confronted with these basic questions about the nature, substance, and funding of the scientific enterprise.

In the United States, many scientists were engaged in a movement to wrest authority over future atomic research and development from the military. At the same time, a movement spearheaded by Vannevar Bush, who had served as director of the Office of Scientific Research and Development (OSRD), sought to secure future government support for basic research.[5] Reform was also on the minds of many young Japanese scientists who saw the ending of the war as an opportunity to break away from the feudalistic traditions that had dominated the scientific community since the Meiji Restoration in the late 1800s. At the same time, Japanese scientists in general, and Japanese nuclear theorists in particular, were eager to return to the community of international science, from which they had been excluded during the war years.[6] But the reconstruction of Japanese science and technology would not be left to the Japanese alone. Like every other Japanese social institution in the immediate postwar years, the scientific establishment would be subject to American influence and control. The orders to MacArthur in this area, however, focused entirely on demilitarization.

Washington policy makers seem to have paid scant attention to science and technology in their planning for the occupation of Japan. The Initial Post-Surrender Policy observed only that "specialized research and instruction directed to the development of war-making power are to be prohibited."[7] The more specific military directive of November 3, 1945 (JCS 1380/15) instructed the occupation authorities to "insure that all laboratories, research institutes, and similar technological organizations are closed immediately except those you deem necessary to the purposes of the occupation." It continued:

You will provide for the maintenance and security of physical facilities thereof when you deem necessary, and for the detention of such personnel as are of interest to your technological or counter-intelligence investigations. You will at once investigate the character of the study and research that have an obviously peaceful purpose under appropriate regulations which (1) define the specific type of research permitted, (2) provide for frequent inspection, (3) require free disclosure to you of the results of the research, and (4) impose severe penalties, including permanent closure of the offending institution whenever the regulations are violated.[8]

Atomic policy was amplified in a directive, issued in December 1945,[9] that provided for the release from custody of Japanese scientists formerly engaged in atomic energy research and the discontinuance of the guarding of Japanese laboratories to which Japanese might have access. Prohibitions on atomic energy research were to be continued, however, and all scientists, instructors, and students familiar with atomic energy research were to be registered. Periodic checks of their activities and of the laboratories in which atomic energy might be studied were also required. These were among the duties assigned to a small group within the ESS known as the Special Projects Unit. While the unit was assigned to the Scientific and Technical Division for administrative purposes, it was made directly responsible to the chief of the ESS, and it reported to him on all matters requiring ESS approval.

The first priority of the occupation forces was demilitarization. Troops were demobilized, munitions plants were closed, and military equipment was seized for reparations. On November 3, Washington directed GHQ to undertake additional steps to shut down research institutes and laboratories.[10] Although the order authorized rapid resumption of research with an "obviously peaceful purpose," it probably influenced the destruction of Japan's cyclotrons.

The Destruction of the Cyclotrons

In between the September 22 and November 3 directives of the Joint Chiefs, Yoshio Nishina, one of Japan's leading nuclear physicists and the director of the Physical and Chemical Research Institute (Rikagaku Kenkyu-jo—widely known as Riken) in Tokyo, filed a request with SCAP for permission to resume the operation of two cyclotrons for research in biology, chemistry, and metallurgy. During the war years, Riken, like many Japanese industrial and research enterprises, had grown with Japan's investment in military research.[11] By the end of the war it employed more than 1,500 researchers in 33 separate laboratories. Much of its work was war-related, from research on cosmic rays to experiments in the Nuclear Research Laboratory; for this reason, Riken was a target of investigation and was subject to the post-surrender policy of shutting down specialized military research and instruction.

Nishina, who had studied in Copenhagen with Niels Bohr, founded the Nuclear Research Laboratory at Riken in 1931. He soon set out to enhance the laboratory's capacity for advanced nuclear research, adding a Wilson cloud chamber, Geiger counters, and a Van de Graaff generator. By 1937 the laboratory had constructed a 23-ton cyclotron, the first to be built and operated successfully outside the United States. The same year, Nishina began work on a 220-ton, 60-inch instrument for the production of neutron sources. It was to be as large as one planned by Ernest Lawrence, who had built the prototype of the first cyclotron at Berkeley. Lawrence had helped leading nuclear research groups around the world develop knowledge about cyclotrons, and he was delighted with Nishina's plans.[12]

Before the advent of nuclear reactors, the cyclotron, which disintegrates atoms by bombarding their nuclei with protons accelerated by electromagnetic force, was the most effective means of producing radioisotopes. Central to research in nuclear physics, the cyclotron produced neutron sources that could be used to create new isotopes.[13] The isotopes produced at Riken, including radioactive carbon, sodium, phosphorous, and copper, were used as tracers in scientific research—especially by the biologists who came to work with Nishina.[14] Eager to resume the scientific research that had been cut short by the war, Nishina immediately pressed the occupation forces for permission to utilize the cyclotrons to produce the much-needed radioisotopes.

During the war, Nishina had been appointed to head a committee, the "Ni (for Nishina) Project," whose goal was to develop an atomic weapon. Benefiting from an influx of government funds through the Japan Society for the Promotion of Scientific Research, Nishina's laboratory grew to approximately 110 scientists by the time the war with the United States began, in 1941.[15] The laboratory's war effort was aimed at developing a method for isolating the U-235 isotope. A uranium isotope was produced; however, for many economic, technical, and material reasons, not the least of which was that Japan had insufficient access to uranium at the time, the project failed. According to Japanese documents analyzed by historians, Japan's effort never even came close to those of the United States, Great Britain, and Germany. The Ni project was effectively abandoned in May 1945.[16]

By war's end, none of Japan's five cyclotrons (two at Riken, two at Osaka University, and one at Kyoto University) was in working order. While Riken's cyclotrons survived the firebombings of Tokyo, in which nearly two-thirds of the institute's buildings were destroyed, there was not enough electricity available to operate them.

In September 1945, Karl Taylor Compton, president of the Massachusetts Institute of Technology, led the first postwar Scientific Intelligence Survey of Japan. Compton recommended that GHQ-SCAP authorize operation of the cyclotrons for the purposes requested by Nishina: experiments in biology and medicine aimed primarily at improving agriculture, forestry, animal husbandry, fisheries, and medical therapy. The permission was granted, but later it was restricted to investigations in biology and medicine.[17]

Even so, by Nishina's account, as laboratory personnel prepared for experiments along the lines authorized, the laboratory was suddenly and without prior notification investigated by SCAP personnel on November 22.[18] Two days later, engineers from the

Eighth and Sixth Armies moved into Riken and the universities at Osaka and Kyoto and began to dismantle Japan's cyclotrons. When Nishina went to the Central Liaison Office of GHQ to protest the action, he was incorrectly told that American scientists (notably Compton) had been consulted in the decision.

Occupation records indicate that the destruction of certain scientific equipment and the seizure of related research data were the responsibilities of the Industrial Division of the Economic and Scientific Section. On November 19, 1945, a top-secret personal directive from the Secretary of War to MacArthur ordered "certain actions . . . which are contrary to the authorization granted to the Japanese government for the use of Dr. Nishina's cyclotron."[19] On November 24, in a memo for the record, the chief of the Industrial Division noted:

Arrangements have now been completed for destruction of five cyclotrons. . . . As of 23 November, all available records have been searched and seized and are in the custody of the sixth and eighth armies. . . . In the course of the above mission, Major O'Hearn (Division Chief) found evidence of other equipment and research relating to atomic energy. These should be investigated and necessary action determined as rapidly as possible, but to do so, appropriate scientific assistance is essential. Such investigations can be accomplished if the two scientists mentioned in the attached radio are dispatched at the earliest date.

Although the "radio" memo requested two scientists from the Manhattan Project, it probably resulted in the assignment of two scientists from the Radiation Laboratory: Gerald Fox and Harry Kelly.

A Scientific and Technical Division existed within ESS at the time; however, it does not seem to have been involved in the investigation or in the destruction of the cyclotrons. At the beginning of the occupation, the division was staffed by a small group of Australians, led by Brigadier John O'Brien, and Australia at that time was not a party to the Atomic Secrecy Agreement. In a memorandum to ESS Chief W. F. Marquat, O'Hearn argued for the creation of a special "Atomic Energy Control Division" within ESS, with himself as head, in place of the Scientific and Technical Division. "Australia," he noted, "is not a party to Atomic Secrecy (U.S., England and Canada only)."

The Response of U.S. Scientists

News of the destruction of the cyclotrons reached the United States through a GHQ-SCAP press release which stated that it had been carried out on orders from Washington. The War Department at first denied having issued any such order, insisting that the order had been for "seizure." The ensuing confusion over responsibility for the action did nothing to relieve the furor that it caused in the United States.

Associations of scientists across the country issued statements and press releases denouncing the destruction.[20] Scientists at Oak Ridge and Los Alamos likened it to the burning of books. Louis Ridenour, a Radiation Laboratory member who headed the Association of Cambridge Scientists, issued a statement that the act implied that "the War Department opposes the advancement of basic science."[21] Secretary of War Robert P. Patterson at first responded to the criticisms by saying that the step had been taken in compliance with the War Department's policy of preventing military research in Japan, but eventually he would acknowledge the Army's mistake.

As other historical accounts of the period have demonstrated, the debate highlighted the sharp disagreement between the scientific community and the military on matters of policy affecting science and technology. It fueled the scientists' argument that military personnel without advanced scientific training were not qualified to control the exchange of scientific information or to make science policy. For example, the Association of Oak Ridge Scientists charged in a press release that "men who cannot distinguish between the usefulness of the research machine and the military importance of a 16-inch gun have no place in positions of authority."[22] According to one historian of the period, the incident demonstrated for many American scientists why science in general and control of the atom in particular should not remain under military control.[23]

In response to criticisms by the leaders of the wartime scientific effort, including Karl Taylor Compton, Ernest D. Lawrence, Vannevar Bush, and Lee DuBridge, Patterson ultimately took responsibility for the order.[24] In a wire to the Association of Oak Ridge Scientists, he acknowledged that "the advice of scientists should have been sought" and said he "regretted the hasty action" of his department.[25]

The military's quasi-admission did little to assuage those scientists who felt strongly that the error should be rectified. On December 31, 1945, Lee DuBridge, the director of the Radiation Laboratory, wrote to Frank Jewett, the president of the National Academy of Sciences:

If [the destruction of the cyclotrons] were indeed a mistake, then it should be rectified. A considerable number of individuals have raised the question with me as to whether it would not be appropriate to raise funds in this country and to collect materials and restore at least Dr. Nishina's 60-inch instrument. The latest discussion of this proposal has come from Ernest Lawrence and Robert Oppenheimer, both of whom are enthusiastically and heartily in favor of such a move. They have requested me to explore opinion in the East on the subject, and my feeling is that physicists in general would favor such a move.

I should like to raise the question with you as to whether you think it would be appropriate, or possible, for the National Academy of Sciences to sponsor a move for raising the funds to restore this cyclotron. Many of us believe it is a most important matter of principle that American scientists could make a very important move toward international good will and the principle of internationalism in science by rectifying the tragic mistake that the War Department has made.[26]

At first sympathetic to the idea, Jewett forwarded the proposal to the Secretary of War. On January 28, 1946, he received a thoughtful response from Acting Secretary of War Kenneth Royall. Royall expressed appreciation for Jewett's courtesy in seeking the War Department's opinion on the proposal, but argued strongly against it:

The destruction of the cyclotrons was a mistake, or more accurately, an accident of war. As such, it was only one of many accidents and I feel that any action to amend it now would be harmful and ill-advised. . . .

To single out this one incident and to proceed as Dr. DuBridge suggests would be unfortunate. It is unsound to even intimate that scientists are citizens of the world alone, are internationalists and not loyal to their native lands and are never willing participants in the ambitions of dictators or tyrants. The evidence to the contrary is too overwhelming for the American public to accept this thesis, for modern war is scientific and technical war in toto. Without the scientist or the technical worker the terrible instruments of destruction of the present day world would not have been possible.[27]

Jewett was persuaded, and he reluctantly set the proposal aside.[28] But the issues it raised, namely the internationalism of science and the relation of scientists to their patrons and to society, remained. At the dawn of the nuclear age, scientists in Japan and in the United States, having served the military objectives of their countries during the war, now faced complex questions about the organization of science in peacetime. In both countries the war had had a profound impact on the government-science relationship—an impact that would directly affect the direction of research and development in the postwar era.

In the immediate circumstances, what the cyclotron incident made clear was the need for intermediaries between scientists and the American military in Japan. The Army issued an urgent request for two senior scientists to join the occupation forces as civilian advisers to General MacArthur. What the Army wanted was expertise about nuclear science coupled with ignorance about the Manhattan Project, so that it could learn about Japan's atomic science capability without revealing U.S. secrets. This seems to be the main reason the Army looked to the Radiation Laboratory for advisers, rather than to Oak Ridge or Los Alamos. There is no record of why the occupation forces chose Harry Kelly and Gerald Fox. When Kelly asked why he and Fox had been recruited for the assignment, his interviewer was blunt: "Well, you had no connection with the Manhattan Project. Nobody can learn any secrets from you. We'd like to know what the Japanese know about the [science behind] the Manhattan Project."[29]

Kelly's Background

During the war, Harry Kelly worked on the development of defensive radar at the Radiation Lab. He knew nothing of the Manhattan Project, and like many Americans he reacted with mixed emotions to the news of the bombing of Hiroshima. Perhaps more than most people, however, scientists realized the profound implications of that event, not only for the war but for the future of the world and the role of science in contributing to that future.[30]

Kelly talked late into the night of August 6 with his friend Scott Buchanan, a philosopher and dean at St. John's College in Annapolis. Buchanan had come to Cambridge to see if he could interest Kelly in taking a position at St. John's after the war.

In a 1970 commencement address at Pembroke (N.C.) State University, Kelly recounted his conversation with Buchanan when they heard the news of the bomb: "After a long stunned silence, my friend said that the drama was not solely in the fearful death of so many people and the vast devastation of so many works of men. The drama also [resided in the fact that all should heed]—not just the scientists and scholars—that mankind could no longer ignore the sociological, economic, and human effects of technology. Although technology was growing silently at an ever accelerating rate, because of the bomb we all awoke to its growth and influence on all men."[31]

What Kelly could not realize then was that his personal history was soon to intersect with events taking shape thousands of miles away in a devastated and defeated Japan. Nor could he imagine then the role he would play in the restructuring of Japan's administration of science and technology—a role based in part on his growing conscious appreciation of the relationship between science and human affairs.

A Biographical Digression

Harry Kelly was born in 1908 in Wilkes-Barre, Pennsylvania. His Irish-American father was a blacksmith for the coal mining industry in the region. His mother, Josephine (née Reilly), cleaned houses to earn extra money for the family. There were three other children: John, Mary, and Catherine.

Harry claimed that his parents' legacy was a lifelong appreciation for honest work and thrift. A good, hard-working student interested in science, Harry went on to study physics at Lehigh University and then at MIT, where he received his Ph.D. in physics in 1936. While at MIT, he became friends with a group of local artists and writers and cultivated his appreciation for the humanities in general and philosophy in particular.[32] Although a scientist himself, Kelly would remain concerned throughout his career about insufficient instruction in the humanities at American colleges and universities.

After MIT, Kelly seemed headed for the private sector, but a personal tragedy pointed him in another, more contemplative direction. Home for a few weeks in the summer of 1936, he went for a swim with his brother in a quarry near their parents' home.

Suddenly his brother called out for help from the middle of the quarry. He couldn't swim back. Harry swam out and struggled to bring him back to shore. Despite his attempts at revival, his brother died in his arms.

The double shock of the loss of his brother and his own inability to save him was a powerful blow to Kelly's spirit. He needed time away from familiar surroundings to think. He needed solace, and he sought it in the vast landscape of the Rocky Mountains, where he accepted a position as associate professor of physics at Montana State College in Bozeman. While there he met and married the college's medical nurse, Irene Andes, the daughter of a Montana homesteader. Prior to taking the position at Montana State, Irene Andes had worked as the sole nurse for Yellowstone National Park and as a visiting public health nurse traveling throughout the state alone by car. She met Harry Kelly when he came into the college infirmary for treatment of an eye infection during his first year of teaching there. Kelly taught physics for four years in Montana. He also developed a technique for using fluorescence to detect disease in potato plants, and he produced a good textbook on electricity and magnetism (published by John Wiley and Sons in 1941). By the summer of 1941, however, he was ready to move back east.

Before September the Kellys were on their way to Annapolis. Harry had accepted a position on the faculty of St. John's College from Scott Buchanan, whom he had met earlier that year at a conference. Buchanan had given a talk on the "Great Books" approach to instruction that had been developed by the St. John's faculty. Students, Buchanan explained, were required to read and discuss 100 classics, many in the original Greek or Latin. The academic climate he described appealed to Kelly, who had considered studying philosophy before he turned to physics. Once they became acquainted, Kelly and Buchanan found they had much in common philosophically and intellectually. When Buchanan later suggested that Kelly come to St. John's to organize the college's laboratories, he accepted without hesitation. Although he stayed at the college less than a year, Kelly would later recall it was one of his most valuable professional experiences. In particular, he reflected later on how his short stint at St. John's had prepared him for his work in Japan which, he claimed, demanded skills and knowledge that extended far beyond the field of physics.[33]

In December 1941, in the wake of Japan's attack on Pearl Harbor and the subsequent declaration of war, Kelly was among the thousands of scientists and engineers from around the United States who were mobilized for war-related research. He accepted a position at the Radiation Laboratory, where he was to meet Gerald Fox, a fellow physicst who had been at the University of Iowa. Launched with a handful of nuclear physicists in a two-room office on the MIT campus in 1940, the "Rad Lab," as it came to be known, grew to become the largest laboratory of its kind in the world, its staff comprising about one-fifth of the nation's physicists.[34]

As the war came to an end, work on radar was curtailed, and the Radiation Laboratory began to close down in 1945. In November, Kelly accepted an offer to teach at his alma mater, Lehigh University. However, in January 1946 he was offered a job with the Occupation Forces in Japan. Kelly never knew for sure why he had been asked, but he was certain that his friend Gerald Fox had been recruited first and had asked for Kelly to accompany him.[35]

Fox and Kelly would be assigned to the Economic and Scientific Section, the largest and perhaps most controversial of the nine civilian administrative offices within the bureaucracy of GHQ-SCAP. Only weeks before the army recruiter came to Cambridge, the section was embroiled in the embarrassment of the destruction of the cyclotrons.

Kelly was less than outraged by the cyclotron incident, viewing it in the context of the many mistakes and cruelties of war. For him, the significance of the cyclotron issue lay less in the loss of the equipment itself than in the dialogue it opened up over the part that science could play in peace and reconstruction and over the appropriate role of science in society.[36] These were the issues that preoccupied Kelly as he prepared to leave for Tokyo in the closing weeks of 1945.

The Departure for Japan

Before they left Cambridge, Kelly and Fox attempted to organize a briefing for themselves by some MIT colleagues who had participated in atomic bomb survey teams, but to little avail. They managed to arrange only a single discussion with one MIT engineer who had been on one of the missions.[37] They received little in the way of a briefing from their recruiter, and no training whatsoever in the language or the customs of Japan.

When Kelly and Fox arrived in Japan, January 2, 1946, they expected their mission to be intelligence-based and to last for about three months. (As it turned out, Fox went home after eight months; Kelly became Deputy Chief of the Scientific and Technical Division and stayed in Japan for four years.)

In addition to his training as a physicist, Kelly brought to his job a humanistic vision of the scientific endeavor that extended beyond national borders and politics. By the time he left, speaking only a smattering of Japanese, he had won the respect and friendship of many Japanese scientists with whom he had worked to reorganize the country's shattered scientific and technological base and to bring Japanese scientists back into the mainstream of the international scientific community.

Industrious, ambitious, and thoughtful, Kelly traveled throughout Japan, going well beyond the limited scope of the intelligence function he had been sent over to pursue. He believed that establishing an enduring democracy in Japan depended at least in part on the country's economic rehabilitation, and that science would necessarily play a key role in bringing about that rehabilitation. At the same time, he found Japan's scientific establishment deeply entrenched in research traditions that overemphasized basic science, with scant attention to the kind of applied research and development that the country needed to meet its immediate economic needs. Many of the young Japanese scientists Kelly met shared his assessment and his desire to reorganize Japan's scientific establishment.

By introducing a policy of mutual trust and by using his resourcefulness to break through bureaucratic barriers, Kelly helped save many scientists and laboratories from punitive action in the early years of the occupation, facilitated the reentry of the Japanese scientific community into the international arena, and helped reorganize Japan's scientific and technological base. For his efforts, he was awarded the first honorary membership ever granted by the Physical Society of Japan, and in 1969 he was awarded the Order of the Sacred Treasure, Second Class—the highest honor the Japanese government bestows on foreigners.

Forces for Change in Japan's Science Policy

2

Kelly Arrives in Tokyo

Harry Kelly's first glimpse of Tokyo, from the window of a plane, was of long stretches of cityscape almost totally flattened except for a few structures poking out of the rubble. It was January 11, 1946, in a cold and harsh winter. Thousands of the million Tokyo residents who had been left homeless by the bombing raids of the previous spring and summer were still building shelters from scrap. Food was scarce. The painful conditions evoked in Kelly a profound sympathy and admiration for the people who seemed to him to be "crawling back on their hands and knees to start building again . . . with tin cans and anything else they could find."[1]

Arriving in Tokyo less than two months after the cyclotron incident and while the controversy still simmered in the United States, Kelly and Fox were greeted with less than VIP treatment. In fact, they were not greeted at all. No one met them at the airport, and they could find no directions to their office. They weren't even sure what office they were to report to.[2] Eventually they found a GHQ-SCAP telephone directory with a listing for an Economic and Scientific Section. Since it had the word *science* in it, they decided "that must be right for us."[3]

At the Dai Ichi Headquarters building in downtown Tokyo they found Major General Marquat, a close confidant of MacArthur who had just been appointed to head the ESS. A career soldier who was probably less at home with science, planning, and economics than with military strategy, Marquat was eager to have technical expertise on board. "And just where the hell have you guys been?" was his greeting to the perplexed newcomers. Marquat was a hard-working and competent administrator who generally focused on broad policy issues, leaving the details of making and implementing policy to his division chiefs. This is not surprising, since the ESS was the largest civilian office in SCAP and had to deal with some of the thorniest issues facing the occupation, including labor relations, reparations, the busting of *zaibatsu* (conglomerates), and economic reform.[4]

The ESS was charged with advising SCAP on economics, industry, and finance as well as science and technology. Over 1,000 people worked there, including Japanese nationals in administrative, clerical, and translating positions. In 1946 the ESS was divided into nine divisions, responsible for cartels and antitrust matters, exports and imports, finance, labor, legal issues, industries, rationing and price controls, science and technology, and statistics and research. Kelly and Fox reported to Australian Brigadier John O'Brien, a personable man who had organized the Scientific and Technical Division into four groups: fundamental research, engineering, chemicals, and general technology.[5] Fox was named science adviser. Kelly was initially assigned to fundamental research but also served in the Special Projects Unit (to address atomic research issues) and as science adviser before being named deputy chief of the division in 1948.

Given their military orders, Kelly and Fox initially assumed that their mission was to be intelligence-oriented and of fairly short duration. Basically, it seemed to Kelly that science advisers were needed because the occupation forces were insecure in distinguishing between military research and appropriate peacetime research. He sensed from the beginning that Marquat didn't really know what specific kind of advice he wanted, only that it should be aimed at keeping the ESS out of the kind of controversy in which it had recently been embroiled. According to Kelly, Marquat's attitude toward him and Fox was pretty much one of "keep them off our backs," but he made it clear that "if we got into trouble, we were going to be sent home on the next boat."[6] How they would carry out the job would be left up to Kelly, Fox, and O'Brien.

They spent the early weeks on the job sorting through bureaucratic details to get some sense of their responsibilities and their authority. Basically, the Scientific and Technical Division served as a coordinating authority in GHQ for all general matters concerning science. Its activities covered a broad range of initiatives, from an abstracting program that employed 20 expert abstractors of scientific papers to documenting all Japanese scientists and scientific societies to promoting the revitalization of Japanese science in Japan and internationally.[7] Kelly plunged into the task with enthusiasm and tenacity, often working 11 to 14 hours a day, 6 days a week, reading reports and meeting with as many people as he could. Increasingly, he felt that the bureaucratic details threatened to frustrate rather than facilitate the evaluation of the state of Japanese science. Early memoranda from Kelly and Fox to O'Brien and Marquat convey their frustration. They complained that "more reports are being written than are justified," that "too many simultaneous investigations are in progress," and that "too many offices are dealing with the same subjects." In response to a long memo from Kelly and Fox proposing a simplified organization for the section, an equally frustrated Marquat sent a handwritten note in response: "A splendid summary of the situation. . . . Keep pressing for final action."[8]

In March 1946, Harry Kelly wrote the following to his wife, Irene:

> I'll try to answer some of your questions. Our job (Gerry and me) is to keep surveillance over research and development in Japan, advise GHQ on policy and action to be taken in regard to all matters pertaining to science and the teaching of science.
>
> Our policy is that all science and technology will be allowed except that directed toward manufacturing of implements of war. . . . Most of our time is spent in changing ridiculous directives already out.[9]

Harry Kelly's first meeting with the Japanese National Research Council's Central Committee for Research, whose membership included a prince, took place on January 13. He listened as the members described what research they had been doing and what they would like to continue to do, aware of the heavy responsibility of deciding what, if any, research was inherently dangerous to security interests. "I find it difficult at present," Harry wrote to Irene, "to squelch any research. It would be so much better if we could come back in 10 years to do our job."[10]

One of the early tasks of the Scientific and Technical Division was to ascertain, in as short a period and to as accurate a degree as possible, the post-surrender condition of Japanese science and technology. An Initial Scientific Intelligence Survey had been made shortly after the September arrival of the occupation forces, in part to ascertain the extent of Japanese research on atomic energy. The survey team, led by Karl Taylor Compton, had revealed the destitute condition of many university and government laboratories; however, the details of the report were considered top secret. It was initially withheld from the Scientific and Technical Division,[11] and it was not available to Kelly and Fox for several months. A major question for them, then, was simply how to proceed.

In keeping with the occupation force's policy of working through the Japanese government, Kelly decided to issue a directive through Japanese authorities stating that he wanted to receive "all secret, top secret, technical, and scientific papers that they had."[12]

The Japanese scientific community complied to the letter with the request, inundating the division with memoranda and reports. Most of the reports were in Japanese and had to be translated for the Americans. They began to pile up, first in Kelly's office and then into the corridors of the Dai Ichi Headquarters. Kelly came to believe that it would take virtually every scientist in the United States to plough through the documentation he was receiving. The insurmountable backlog created an impossible situation for Kelly and Fox. Kelly's initial assessment and his conversations with Japanese and American scientists had by this time convinced him that the scientific enterprise posed no immediate military threat, and it was clear that collecting written reports would not substantially improve the intelligence function. Rather than continue going through the reports in Tokyo, Kelly decided to forward them to the Joint Chiefs of Staff in Washington. He would seek other avenues to fulfill the intelligence responsibility and to begin the work of reconstruction.

In order to be an effective intermediary for the reconstruction of Japanese science, Kelly needed to have credibility with both the Japanese scientific community and the military command of the occupation. This, in his view, required far more than field inspections and written reports from the nearly 1,300 laboratories that the division had documented. It meant that both sides would have to trust him. The Japanese would have to trust his motivations, and the Americans would have to trust his ability to know what was going on in Japanese science. The solution (proposed, according to Kelly, by the Japanese) was, essentially, self-policing. Acting as Chief of the Fundamental Research Branch, with Fox as Science Adviser, Kelly developed an overall policy that permitted "research and teaching for the extension of scientific and technical knowledge except activities directed toward developments in the field of warlike activities with specific restrictions in aeronautics and nuclear physics."[13] Understanding of the research was maintained through self-reporting and routine field investigations. "The Japanese . . . [understood] our position," Kelly wrote, "and would report to us any questionable research or development activity. During my whole stay in Japan I never found a shred of evidence to suggest that this promise was violated."[14]

A Good Beginning

One incident that occurred early in Kelly's tenure in Japan served as an inspiration for the cooperative approach that would become the hallmark of his efforts. It took place during a site visit to Hokkaido University, where Kelly, accompanied by a chemist from the Scientific and Technical Division and two interpreters, went to investigate rumors of "death ray" research at the Microwave Research Institute. What was said to be going on was an experiment to determine whether high-energy electromagnetic waves in the microwave region could melt metal when aimed at an aircraft.[15]

Located outside Sapporo on the northernmost of Japan's main islands, the Hokkaido University campus in February 1946 was a desolate place that mirrored conditions at many of Japan's universities in the first postwar winter. Classrooms were deserted. University personnel kept small stoves operating in the research laboratories just to warm themselves in the freezing rooms. Little work was underway. The professors and other personnel at the university were preoccupied with fending off starvation and cold and dealing with rumors of reprisals by the occupation forces.

Reports of the Riken cyclotrons' being thrown into Tokyo Bay had reached the Hokkaido campus, along with stories about the requisitioning of buildings for occupation use, orders to purge the universities, and reports of occupation forces ransacking research facilities in Tokyo. Before Kelly arrived, a local edition of the Tokyo daily newspaper *Mainichi* reported that three Hokkaido professors—

Takashi Minoshima, Ukichiro Nakaya, and Yoshihiro Asami—were suspected of war crimes. Minoshima was head of the laboratory Kelly was supposed to investigate.[16] When word came that this representative from SCAP headquarters was on his way, the faculty feared the worst.

Kelly was apparently unaware of the rumored war-crimes charges, and he went about the business of inspecting the laboratories in a matter-of-fact way. No one mentioned dismantling or requisitioning research facilities for other purposes. No one asked to interrogate any professors. No one was arrested. Yet Kelly was not completely satisfied. The laboratories were too empty, and he sensed that something was amiss. He reported this to the university authorities and asked where the missing equipment was. His request was met with silence, and he spent several fruitless days on the campus waiting for someone to come forward.[17] When he was about to return to Tokyo, he left word for university officials that he still believed some apparatus was missing and that he would like to know more about it before leaving.

Kelly's note caused a stir among some of the faculty. They decided to send a small delegation to speak with him alone, as a scientist rather than as a representative of the occupation.

Among the delegates was chemistry professor Juro Horiuchi. An outspoken intellectual, Horiuchi was eager to talk to Kelly, not because he had something on military research to report but because he was angry. He was in fact outraged by the quartering of American troops in the metallurgy laboratory and the Institute of Low Temperature Science. He confronted Kelly with stories about the soldiers' unsavory behavior and challenged him to do something about it. Expecting at best a lukewarm response, Horiuchi was surprised by Kelly's immediate assurance that he would help to rectify the situation. This opened up a longer conversation between the two men about the conditions for research and the organization of Japanese science. Horiuchi was completely candid about his feelings that the existing framework for contact among Japanese academic scientists was in fact an impediment to scientific development. In particular, the Japan Academy, according to Horiuchi, functioned solely as an honor-conferring institution and played no constructive role whatsoever in promoting Japanese science. What was needed, according to the feisty professor, was an organization of energetic scholars who would be active on both the Japanese and the international scene.

Kelly expressed his appreciation for Horiuchi's candor, complained that he could never get a true sense of conditions from dealing with scientists through the Ministry of Education, and said that he wanted to talk with others who were willing to speak freely.

The communication channels Kelly was expected to use were formal and roundabout. Under the occupation policy of working through the Japanese government, Kelly's most direct link with the scientific community was through the Ministry of Education, which controlled government funds for scientific research. Through the ministry he had contact with the three organizations for science that had emerged from the war with active memberships: the National Research Council (NRC), the Imperial Academy, and the Japan Society for the Promotion of Science (JSPS). These hierarchical bodies were, however, far from representative of the full Japanese scientific community. The number of Japanese scientists engaged in active research after the war was about 25,000, with nearly half of those in engineering. Membership in learned societies totaled 414,000, of whom 257,000 were engineers. Because of the scarcity of resources during the war, many of the learned societies had ceased to meet or to publish papers, and by 1947 only about 50 of them claimed memberships exceeding 500.[18]

Many scientists, like Horiuchi, emerged from the war with the strong conviction that the NRC, the JSPS, and the Imperial Academy were too feudalistically inclined, too dependent for advice and research proposals on scientists in the Tokyo area, and too tainted by war-related activities to serve as appropriate vehicles for reviving Japanese scientific efforts.[19] The NRC had played a significant role in the Japanese government's efforts to direct the activities of Japan's scientists toward military research and development. Although largely unsuccessful, these R&D efforts had increased the NRC's membership and prestige[20] at the cost of an association with the militarism that many scientists were turning against in the wake of the defeat. In addition to their criticism of those who had led the country into war, many young Japanese scientists were critical of the Imperial Academy, a primarily honorary body whose elderly members no longer contributed actively to their fields but nonetheless prevented the younger scientists from advancing their opinions about support for scientific research.[21]

This dissatisfaction and desire for change manifested itself in demands by individuals and organized groups for reforms ranging from minor changes in the rules of membership in the Imperial Academy to breaking with militarism and embarking on research pursuits aimed at promoting peace.[22] Because the occupation aimed at fundamental institutional reform, the Scientific and Technical Division would inevitably become involved in and play a crucial role in this debate.

Kelly's meeting with Horiuchi was probably his first informal introduction to these debates, and he was energized by it. Writing to Irene from Hokkaido on February 21, he said:

Golly, I'm busy, but I enjoy it very much. . . . Such confusion—each dept. at the Univ. is a little Kingdom which doesn't know what the other dept. is doing. They haven't started school. Maybe they will before we leave —we're doing the President's work and bawling everybody out. . . .

We succeeded in convincing the army that they shouldn't take anymore buildings on the campus. This proved to the college that we were really sincere when we said we'd help them and they opened up and gave us much more information than we expected.[23]

Horiuchi was pleased to have found a potential ally. Encouraged by Kelly's surprising agreement with his views, he promised to introduce the American to scientists of real vision, not merely academic "bigwigs." The evening before Kelly's planned departure from Hokkaido, the president of the university, along with several scientists on the faculty, asked him to stay another day. They did, after all, have something they wanted to show him.[24]

In 1945, near the end of the war, some of the university's research equipment (a large magnetron and other electronic gear) had been evacuated to a grade school in Tohbetsu, a small village about 30 miles from Sapporo. Fearing reprisals for engaging in war-related research, the scientists were too frightened to tell anyone what had happened, but Kelly persuaded them that it would be better for them if he could make a full report on the matter.[25]

The next day, Professors Asami and Minoshima accompanied Kelly and a military officer by train to a wooden schoolhouse in the village of Tohbetsu, where they found some of the "death ray" equipment stacked in a corner. Although the research had been aimed at melting airplanes, it was clear to Kelly that the prototypes in the schoolhouse, with a maximum range of 20 feet at best, would never be effective for military purposes. After a brief discussion with the Hokkaido scientists, Kelly decided to spare this equipment so that it could be used for other purposes in the university's labs.

Some tangible gesture was needed to quell the scientists' anxiety that after Kelly's departure military personnel would find the prototypes, misinterpret their potential, and then destroy them or punish the scientists for keeping them. Kelly wasn't quite sure how to handle the situation officially. Taking the most direct route, he hung a large sign over the gear: "Property of Hokkaido University, Electronic Laboratory, inspected by H. C. Kelly, General Headquarters." The stunning contrast between this enlightened approach and what the scientists thought would be standard U.S. military procedure after the cyclotron incident awakened in the Japanese a renewed sense of hope and possibility.

The Hokkaido meetings over, Kelly was eager to leave Tohbetsu for Tokyo. But a blizzard had disrupted train service out of the village, which lacked a hotel or inn. Village officials conferred and decided to house the mission with a prominent local family. By nightfall, the storm had caused a complete power failure that plunged the village into total darkness. Minoshima and Asami brought batteries and candles from their schoolhouse laboratory to provide a dim light for the visitors, but in the cold and dark apprehensions soon surfaced. Kelly later recalled that the military attaché who accompanied the scientific mission worried because he was unarmed. They soon realized, however, that their hosts were even more concerned than they that something bad might happen to them.[26]

The storm persisted, and after two days there was little hope that rail connections would soon be reinstated. Kelly, fearing that the mission would "eat all the food in the village," insisted that the party leave, by sled if need be.[27] The next morning the villagers outfitted two horse-drawn sleighs with blankets and charcoal heaters and sent the five-member party out across the Ishikari Plain to Ebetsu, a journey of about ten miles. They had heard that the trains were running from there to Sapporo.

The adventure was still not over. About five miles out, the sled carrying Asami and Kelly overturned, and they were thrown headlong into the snow. Asami later recalled imagining Kelly's anxiety: to be snowbound in a country that had only recently ceased to be an enemy, stuck in what was probably the worst snowstorm he had ever encountered, and then hurled from a sled into the raging snow must have been a frightening experience indeed!

In fact, Kelly remembered the Hokkaido trip as a positive turning point. In an article for the American Association for the Advancement of Science, he described the candor of the meetings, the hospitality of his hosts, and the sleighing mishap he had shared with his Japanese colleagues: "I was in Japan long enough to know that it is un-Japanese to be profuse in expressing thanks. But I cannot resist the temptation to note that we have as much to learn as we have to teach—even during a military occupation."[28]

Forging Ties: The "Three Musketeers"

Soon after Kelly's departure, Horiuchi began to tell colleagues in Tokyo that a fellow scientist was working for the occupation and that he was someone they could work with.[29] One of these colleagues was his close friend Horishi Tamiya, a plant physiologist at the University of Tokyo. Horiuchi and Tamiya agreed to meet in Tokyo in the spring of 1946. Reluctant at first, Tamiya ultimately agreed to go along with Horiuchi's plan to discuss the reorganization of Japanese science with Kelly at occupation headquarters.

Born in 1903, Tamiya studied plant physiology at the University of Tokyo. He worked his way up the faculty ranks from junior assistant to instructor prior to being sent abroad to study in Europe and the United States. He returned to Japan in 1939 and soon thereafter was named a professor at Tokyo University. A prominent scientist in his field, Tamiya won international recognition for his research on photosynthesis and on plant and cell physiology as well as in the field of microbiology. He played a key role in rebuilding the university after the war and was named director of its Applied Microbiology Research Institute. Later, he became director of the Tokugawa Biological Research Institute.

After the war, Tamiya lived with his family in a room in one of the university's science faculty buildings. When Horiuchi was in town, the two scientists often met there, talking and arguing late into the night. When they talked about what had happened to their country, Horiuchi vigorously denounced Japan's former enemies, while Tamiya blamed Japan's leaders for the defeat. Their friendship was unusual. The two men practiced in different fields, lived hundreds of miles apart, and had completely different personalities. But they agreed on one thing: the need for change in Japan's science organization.

The initial conversation at occupation headquarters in the spring of 1946 established a pattern of frankness that would continue throughout years of friendship. Speaking candidly, Tamiya told Kelly that the blunt reform approach proposed by Kelly and Horiuchi was "too simple-minded." He advocated a more carefully planned process, one that began with an open forum of scientists from all fields to discuss current aspects of the Japanese scientific research system and the specific problems affecting each individual discipline. A necessary first step, according to Tamiya, would be the organization of a conference to launch the initiative. Kelly agreed. "You're right," he said, "so you must organize the conference. I leave everything to you."

Intimidated at first by the scope of the task, Tamiya later recalled that Kelly induced him to participate by offering him one rare American cigarette after another.[30]

When he left the headquarters, Tamiya had serious misgivings about being able to assemble the meeting with no money and with the extremely limited resources available at the university. He needed help. Horiuchi recommended Seiji Kaya, a physicist who had occupied the laboratory next to Horiuchi's at Hokkaido University before his recent move to the University of Tokyo.

Slightly older than the others, Seiji Kaya was born in 1898 in Kanagawa Prefecture, near Tokyo. After graduating from the physics department of Tohoku University in 1923, he went to work as an assistant to Kotaro Honda, chairman and research director of physics in the faculty of science. Well known for his research in steel, Professor Honda had helped to launch the Research Institute for Iron and Steel at Tohoku University. Like Japan's Physical and Chemical Research Institute, this institute was established with private donations from major industrialists to promote self-sufficiency after the curtailment of imports during World War I.[31] In 1922 the facility was nationalized as the Research Institute for

Iron, Steel, and Other Metals, and it was there that Kaya worked as Honda's assistant.

In 1928, Kaya left Tohoku to study in the United States and in Europe (where he spent a short time at Berlin's Physical Technology Institute). Upon his return to Japan in 1930, he was appointed assistant professor in the newly established Science Faculty of Hokkaido University. Promoted to full professor within a year, Kaya spent the next 12 years working on the question of a mechanism for magnetization. It was during this period that he and Horiuchi had adjoining laboratories.

In 1942 Kaya was awarded the prestigious Japan Academy Prize for his work on the magnetism of ferromagnetic crystals, and the following year he was appointed a full professor in the Science Faculty of the University of Tokyo. Concurrently, he held a position in the university's Aeronautics Research Institute, where his wartime research focused on alloys and the development of radar weapons.

Horiuchi saw in Kaya an opportunity to break the hold of the old elite and to inject "new blood" into the science establishment. Kaya had the right Tokyo credentials to be accepted by the establishment; at the same time, as a graduate of Tohoku University and a former professor at Hokkaido, he represented a real break from Tokyo's deeply entrenched "old boy network." Even more important to Horiuchi was what he referred to as Kaya's "rustic" personality. Walking over to Kaya's university laboratory with Tamiya, Horiuchi told him that the job of renewing Japanese science would be too tough to leave to "weak-nerved Tokyoites." It would take the nerve of a rustic like Kaya, a person who did not get lost in trifling details but would instead focus on broader goals. Although Tamiya was a product of elite Tokyo schools and could be viewed as just the sort of thoroughbred Tokyoite that Horiuchi worried about, he was viewed as something of an outsider within the Tokyo-based science establishment.[32]

Kaya was a pragmatist who wanted Japanese scientists to have a say in government policy.[33] He agreed to work with Kelly and the other scientists, and he asked to bring in another University of Tokyo physicist, Ryokichi Sagane.

Sagane was the youngest son of Hantaro Nagaoka, a pioneer in Japanese physics.[34] Perhaps adopted, he was the only one of Nagaoka's five sons to follow in his father's footsteps; the others all pursued careers in engineering. Nagaoka was one of Japan's most prestigious scientists. In 1917, together with Masatochi Okochi, he had founded the physics section of the Physical and Chemical Research Institute (Riken). In 1946, at the age of 80, he was the head of the Imperial Academy, which his son now challenged as the most important advisory body on Japanese science.

Sagane, after his graduation from the University of Tokyo in 1929, worked in Riken. In 1931 he began to spend part of his time in Nishina's newly established laboratory there. Between 1935 and 1938, like many of Japan's most promising physicists, he was sent abroad to study nuclear physics and cosmic rays in Europe and the United States. Particularly gifted, Sagane studied with the world's leading physicists, including Niels Bohr in Copenhagen and Ernest Lawrence at Berkeley. At Nishina's request, Sagane was invited by Lawrence to be part of the Berkeley cyclotron team, along with about ten other visiting physicists. During his stay in the United States, Sagane visited nearly all the cyclotron installations at universities, met many American physicists, and became an expert on cyclotrons.[35] Upon returning to Japan, he completed his doctoral thesis on "artificial radioactivity." He received his doctorate in 1939, when he was already an assistant professor at the University of Tokyo and a researcher at Riken. As a key member of Nishina's experimental group on nuclear science, Sagane played a major role in the construction of the Riken cyclotrons. In 1946, when Kelly met him, he was a full professor at the University of Tokyo.

Close in age (Sagane was then 41, Kelly 38), Sagane and Kelly were both strong-willed men of vision. They understood each other and got along well. Many years later, Kaneshige Kankuro, who served as Chairman of the Scientific Research System Renewal Committee, remembered impromptu meetings between Sagane and Kelly. In a television documentary he recalled:

After meetings of the Committee, I used to go and see Dr. Kelly to give an account of our deliberations. Not being proficient in English, I asked Dr. Sagane to help me out with the language. The way he did this job for me reminded me of a story that is told about a Japanese mission that went to Europe at the time of the Meiji Restoration. The delegates had previously briefed their interpreter on what they wished to say. When time came to deliver the statement, the chief delegate just said a few words to set the interpreter on his track, and then left him to deliver the rest. It is said that their foreign partners in the dialogue were astounded at the amount of meaning that appeared to be compressible into a tiny bit of speech in Japanese. Well, my dialogue with Kelly sounded a little like that when Sagane took it upon himself to interpret for me.[36]

In fact, Sagane's facility with English and his well-known erudition were indispensable qualities in building mutual respect between SCAP officials and the Japanese scientists. Each of the three members of the group initiated by Horiuchi to work with the Americans brought important strengths to the table. Kaya's personality and prestige among Japanese scientists lent credibility to their efforts, while Tamiya's boundless energy and attention to detail helped them lay a sturdy foundation for change. Kelly called them "The Three Musketeers."

Each of the members of this original group became prominent in his field. Horiuchi eventually became president of Hokkaido University; Kaya became president of the University of Tokyo, the first president who was not one of the school's graduates. Sagane left Japan for the United States in December 1949, but he returned in the 1950s to spearhead nuclear energy research in Japan. He became vice president of the Atomic Energy Research Institute and, later, vice president of the Japan Atomic Power Company.

Beyond Militarism: The Science Liaison Group

Tamiya, Kaya, and Sagane formed the nucleus of a group that laid the groundwork for the creation of the Science Council of Japan. This was not Kelly's original objective, though. As he envisioned it, the group would serve primarily as a link between the occupation forces and the Japanese scientific community to aid in the implementation of occupation policies.

At GHQ meetings, Kelly proposed the creation of a "Scientific Liaison Section within the Japanese Central Liaison office." The primary function of this organization of physicists, chemists, and other scientists would be to gather specific information as required, to help interpret occupation directives to the Japanese, and, by more intimate contact, to bring Japanese problems to the attention of the Americans.[37]

It was with this liaison function in mind that Kelly summoned Tamiya two or three days after Horiuchi had introduced them to ask him to submit the names of about twenty active Japanese scholars in the natural sciences within two weeks. Tamiya was exasperated by the request and offended by Kelly's abrupt and impatient manner. "Please wait a minute," he told the American. He wanted Kelly to try to understand the situation from a Japanese point of view. "Suppose that Japan had won the war with your country, and I was assigned to Washington, to sit as you are sitting here today. Suppose I summoned you before me and demanded the immediate submission of the names of prominent American scientists. What would you do?"[38] Kelly was upset by the implication that his reasons for soliciting the names of scientists were ambiguous and by what he considered a brash response. Years later, however, he told Tamiya: "I was truly vexed at your 'suppose Japan had won the war' comment. . . . But turning it over in my mind, I reckoned you were right. You gave me food for thought. It made me like you, and trust you."[39]

To Kelly's Japanese friends this characteristic honesty that allowed him to rethink issues compensated for the outbursts of temper to which they soon became accustomed. Rather than merely submit a list of names to Kelly, Tamiya set to work to organize a forum for Japanese scholars—primarily scientists—to discuss representation with the occupation forces. Tamiya, who saw in the creation of a Liaison Group an opportunity to effect permanent change in the organization of Japanese science, proceeded cautiously, intent on building lasting support.

On May 22, 1946, a meeting was organized through the Central Liaison Office of the Occupation Headquarters with representatives of the Ministry of Education, the National Research Council, three scientists from the University of Tokyo (Tamiya, Sagane, and Kameyama), one representative from Osaka University, and Yoshio Nishina of Riken. The Ministry of Education agreed to provide funds for the operating expenses of the group, which would start off with 49 representatives—seven from each of seven regions of Japan. At the first meeting of the Japan Association of Science Liaison, on June 6, Nishina argued that if the group was to be considered representative of the Japanese scientific community, the 49 members should be selected through "a public election, after wide consultation with all the universities and official and civil research organizations." Others disagreed, expressing concern that an election under the current impoverished conditions of the scientific community would not necessarily ensure the desired democratic outcome and would be difficult to organize, thus delaying implementation of the liaison group. The latter view prevailed, and it was decided that members would be invited by the groups at the preparatory meeting, namely, the Ministry of Education, the Central Liaison Office, the NRC, Nishina, and representatives from the universities of Tokyo and Osaka, including Kelly's contacts, Tamiya and Sagane.[40]

The first meeting of the Science Liaison Group (Kagaku Shogai Renraku-kai) took place in the Imperial Academy Building, located on the grounds of the Tokyo Science Museum, on July 9, 1946—six months after Kelly's arrival. As Kelly explained, it was designed to be of mutual benefit to the occupation authorities and the Japanese scientists.[41] "My sole gain [as a representative of the occupation]," Kelly said, "is that we [the United States] get a friendly neighbor [in] Japan."[42] He saw the underlying reward for the Japanese to be the restoration of their country's economic health and its international stature.

At the meeting, Kelly acknowledged the formidable obstacles as well as the delicate policy issues that had to be overcome in order to orient the scientific research agenda toward rebuilding the economy as rapidly as possible. Not the least of these were the bureaucratic issues surrounding the administration of resources for science and technology at the government level. The Ministry of Education controlled the distribution of government funds for scientific research and development, and in the Division's assessment this resulted in a continuing overemphasis on fundamental research. Research laboratories under the administrative responsibility of another government agency, the Ministry of Commerce and Industry (forerunner of MITI, the Ministry of International Trade and Industry), were not coordinated. Beyond that, at the inter-ministerial level, no agency was responsible for coordinating overlapping scientific and technical matters throughout the government.[43]

Kelly recognized that only changes that responded to needs identified by the Japanese and were suitable to the Japanese situation could be expected to succeed. "Now is the time to develop an ideal system," he said. "You must not miss this chance. Don't take too long doing this, but on the other hand, don't rush things to a hasty conclusion. And remember, this is your own problem; it must be solved by you for yourselves."[44] Concerned about bureaucratic infighting, he warned about the need to rise above personal interests when addressing questions such as where to locate the nation's leading science advisory group: "Should it be under the [Education] Ministry, the Prime Minister's office, or the Ministry of Commerce and Industry? I don't know. I need advice, Japan needs advice, and you are the men to help give that advice."[45]

At first, scientists from a broad range of fields (including engineering, agriculture, and medicine) joined the Science Liaison Group, but they soon asked permission of SCAP to form independent groups. In each field, the groups appealed especially to active young scientists who found in them a forum for expression that had previously been denied to them by the existing hierarchical academic associations.[46]

Because Kelly was a physicist, the members of the SL Group decided that their primary contact with him should also be a physicist. They selected Seiji Kaya, and it was to him that Kelly formally conveyed changes in occupation policy concerning research prohibitions and other matters.

From the beginning, however, it was clear that the primary function of the group extended far beyond these limited purposes to the actual reorganization of science. Discussions often focused on the ineffectiveness of the major academic associations for science. In agreement with many of the young scientists with whom he met, Kelly concluded that Japan's three existing learned societies for science were more impediments than facilitators of what he felt were the most important roles for science and technology in Japan at that time, namely economic recovery and the promotion of world peace.

Although Kelly recognized the importance of structural change's coming from the Japanese themselves, he nonetheless had strong views of what should happen, and he worked behind the scenes with Tamiya to gain credibility for the SL Group and to use it to advance the goals of the scientists seeking reform. In a meeting on June 22 the two men argued about the appropriate role of the group. Kelly insisted that it should take a proactive role in science reorganization, without waiting for change from within the traditional bodies. He suggested that Tamiya seek assistance from American scientists. "Write a letter from the Japan Association of Science Liaison to the National Research Council in America," he said.[47]

The purpose of the letter would be to explain the situation in Japan and ask for their help. Kelly suggested that Tamiya explain the functions of the new Japanese organization, mention its problems, and describe the kind of contact the Japanese scientists would like to have with the U.S. National Research Council and the kind of help they needed from American scientists. "I don't think we in America appreciate the problems here," he said. "If we could get some indication of the eagerness of the Japanese scientists to help your country and cooperate, I think American scientists would probably be in a much better mood to help out. They know nothing except what is read in the papers. I will go back and see the NRC and ask them if they will help."[48]

Tamiya agreed that seeking help from the Americans was a good idea, but he was uncomfortable with suggesting that the Science Liaison Group represented Japanese scientists. "The Japan NRC and Imperial Academy of Science are representative of Japan science," he said.[49] In the crisp exchange that followed, Kelly insisted that the NRC and the Academy were "not doing the job": "I don't think they are doing very much good. . . . I think the time has come when we have to come out in the open—we are trying to be that representative. That is really what we are striving to do. . . . Those organizations . . . need reorganization. Therefore, it is up to your group to help reorganize this group." Tamiya responded: "Our present group is far from [being representative]. We have the hope, confidence and worth to be called active scientists, but [in] such a political situation, it is not yet time to express this thought. It is too idealistic." Kelly seemed unable to hear Tamiya's argument concerning the political sensitivities involved. "If anything is good," he replied sharply, "it is not too idealistic."[50]

In the end, the two scientists agreed that enlisting the assistance of American scientists was a good idea, that reorganization was necessary and could be carried out even though the task was more difficult from the Japanese perspective than from Kelly's.[51]

The next day, Tamiya delivered to Kelly the draft of a letter addressed to the National Research Council of the United States. Signed by Kaya, Sagane, Tamiya, and 17 other members of the Japan Association for Science Liaison, the letter was sent on July 11. In it Tamiya spelled out the functions and mission of the association, including its desire to contribute to the rehabilitation of Japan and to improve the traditional systems for the organization of science. He pointed to the scientists' dissatisfaction with "the past impotency and clumsiness of our government authorities in utilizing and respecting scientific talents," as well as to the shortcomings of academic scientists' penchant for what he called "academic fogydom."[52] He spelled out the deplorable material conditions under which the Japanese scientific community was struggling to resuscitate its research efforts and to focus those efforts on civilian pursuits to improve the standard of living. "Considering the present state of our scientific and technical world, we are at present restricting our problems to the affairs in the domain of fundamental researches in physics, chemistry, biology and engineering, although we hope to extend our activity step by step to all branches of science and technology in the future."[53]

In its sharp criticism of Japan's traditional power structure, its condemnation of the "thoughtless and erroneous war" into which these powers had led the nation, and its aspirations for a "new Japan which will contribute to the World's Peace and Humanity," Tamiya's letter reflected the views of many progressive Japanese intellectuals in the immediate postwar period. They believed that in the new democratic postwar order Japan would demonstrate that a modern industrial nation could exist without arming itself. This view had been given expression in the country's newly adopted constitution—particularly in article 9, which renounced all war-making activity.[54] In so doing, the constitution definitively set Japanese research and development on a trajectory toward peaceful (that is, civilian) pursuits and away from a military, defense-related course.

The constitution also solved a dilemma that the reorganization of the national scientific bodies posed to the occupation authorities. As Kelly explained in a 1949 article for the journal *Scientific Monthly:* "On one hand was the policy that science and technology were to be encouraged, since they were considered essential to economic recovery; on the other, a powerful governmental scientific body might well serve as the genesis of a new technical General Staff with warlike aims. It was finally decided that if, under the new Constitution, an elected nongovernmental council of scientists were formed, it would be most likely to accept its responsibilities in promoting world peace."[55]

By the middle of 1946 Kelly's mission had clearly changed from surveillance to mediation and friendly guidance. This shift reflected his personal desire to influence the direction of Japan's research and development goals.[56] Kelly's response to the poor living conditions in Japan and his strong belief in the duty of scientists to contribute to human welfare echo throughout his correspondence and his contributions to meetings during this period. He often expressed his sympathy for the Japanese people's economic suffering and his belief that a shift from basic research to applied and developmental science was crucial to the country's future.[57]

Kelly's belief that economic self-sufficiency was crucial to Japan's reemergence as a friendly and prosperous nation came far in advance of the so-called reverse in the course of U.S. policy. (The occupation bureaucracy did not formulate a policy to rebuild the economy of Japan until January 1948, when the Far East Commission announced a "New American Policy for Japan, one that favored economic reconstruction over institutional reform."[58])

In May of 1946, Harry Kelly was formally asked to extend his stay. He and Irene had been considering this possibility for months, and in June he wrote to her as follows: "Today I signed an agreement to stay over here for one year from the time you arrive. Two reasons for the decision. 1) I think you will enjoy it and it will be good for both our educations and 2) I believe I can do some good here. I haven't all the details yet—I'm sure you won't leave before September. Don't be disgusted at army red tape."[59]

In fact, Irene and the Kellys' young son Henry did not get to Tokyo until the following February. Before they arrived, Harry Kelly rented a house in the Setagaya Ward in Tokyo, one of the few that had been spared in the firebombings. "The landlord is a typical Japanese character," Kelly complained, "most ticklish to deal with."

With the decision to stay made and the priorities of the Scientific and Technical Division established, Kelly undertook his tasks with vigor. His colleague Bowen Dees remembered that "Harry had a strange power of intuitive discernment: he would in an instant grasp the essence of a matter. This is a precious quality for a man in a position to command. Many in GHQ—and in any administrative office—were of bureaucratic mind, and would follow to the letter what was said in instructions and orders, but Harry was not that type: he followed his own principles. He was a man true to his convictions."[60] These were qualities that would serve Kelly well in the years ahead.

A U.S. Model for Japan?

3

Toward a Renewal of Japanese Science

The problem of reorganizing science for the postwar world was universal to the industrialized states that had fought in World War II. German science, which until the war had led the world in theoretical research, had collapsed under Nazi control, and many scientists had fled the country. Japanese scientific research and development had also fallen far behind, owing to the isolation of scientists from the international community during the war and the extensive damage to their facilities.

While Harry Kelly was serving as an adviser to the Japanese scientific community, American scientists were going through an organizational crisis of their own. The crucial question concerned the postwar relationship between science and government. The debate within the scientific community focused on two major issues: the struggle for civilian control of atomic energy[1] and the future of basic research. Of particular concern in both cases was the intersection of two potentially incompatible goals: maintaining government support for the scientific enterprise and keeping scientific research insulated from government control and protected from politics.[2] These concerns were at the heart of legislative debates shortly after the war concerning the creation of the National Science Foundation and the Atomic Energy Commission.

Vannevar Bush, who had headed the wartime Office of Scientific Research and Development, worried particularly about the future of basic research.[3] At the request of President Franklin D. Roosevelt, he prepared his now-famous report *Science—the Endless Frontier,* in which he argued for a sustained relationship between government and science.[4] Recommending the creation of a "National Research Foundation" to serve as a focal point for the support and encouragement of basic research and education in the sciences and for the development of national science policy, Bush noted that, while science is a proper concern of government, the U.S. government had only just begun to utilize science in the nation's welfare. The report, submitted in 1945, launched a legislative debate that continued until 1950, when a compromise led to the creation of the National Science Foundation.

In both Japan and the United States, whether for good or ill, the massive commitment of government resources to scientific and technological development during the war and the terrible proof of the effectiveness of this strategy had irrevocably changed the environment for scientific research. Now the problem was how to interpret this experience in planning for the future.

This issue was particularly salient in Japan, where the traditional forums for discussion of key issues were widely regarded as insufficient for the task at hand. Honorary societies and other institutions steeped in hierarchy, it was believed, simply could not respond to the demands that society now placed on science and technology. What was needed, according to those calling for reform, was a forum where energetic scholars could be actively involved in planning for investment in the scientific enterprise and in meeting the needs of society.

Both Kelly and Horiuchi saw a model for such a forum in the U.S. National Research Council (NRC), which was created as a "working arm" of the National Academy of Sciences (NAS). Established in 1863 by congressional charter, the NAS, like the Japanese Imperial Academy, sat at the honorary apex of science, and its membership included the most professionally distinguished scientists in the national community.[5] Although it had originally been intended to serve in an advisory as well as an honorary capacity, its advice was rarely sought. By the time of the First World War, when it became apparent to the NAS that its honorary and advisory functions were not compatible, it set up a separate branch, the NRC, that could draw upon the scientific community at large to respond to issues of public importance. Capable and active scholars from throughout the country were asked to serve on NRC committees to provide analysis and advice, whether or not they were members of the NAS. Kelly, Horiuchi, and others felt that Japan could use a similar advisory body that would be distinct from the honor-conferring academies.

As the work of the Science Liaison Group got underway in Japan, Kelly pursued the possibility of getting American scientists to visit. He wanted "the cream of the American scientific community" to share their insights concerning the American system, drawing what lessons they could from their own experience for their Japanese colleagues and providing advice to the occupation authorities on the stance they should take toward plans proposed by the Japanese for the reorganization of science.

Kelly's office informed the newly created SL Group that occupation headquarters required three operating principles for authorization of the liaison group:

- It must act as an independent body with complete freedom from all branches of the Japanese government.
- It must be constituted of representative, actively engaged scientists.
- Its membership must encompass the whole nation.

These requirements were in keeping with the Economic and Scientific Section's three-part policy for scientific and technical affairs:

democratize Japanese institutions, foster the development of Japanese science, and provide indirect guidance on implementation of GHQ policies. They also reflected Kelly's personal view that direct communications were preferable to reports filtered through official channels.

In addition to the problems of reorganization and the shift in research priorities, SCAP documents reveal a number of other hurdles the Scientific and Technical Division believed had to be overcome in order for Japanese science and technology to develop rapidly. These included the isolation of Japanese scientists during the war, the general impoverishment of the learned societies (including the suspension of publication of many scientific journals even before the end of the war, owing to paper shortages), and the lack of standardization within industry. The Division also suggested revision of Japanese patent laws and their administration, both to improve the protection of imported technologies and to spur indigenous innovation.[6]

The Scientific and Technical Division reported that, because of poor conditions and intellectual isolation, many Japanese industrial plants were obsolete at the start of the occupation. Whereas plants in other industrialized countries were starting to use accumulated wartime scientific discoveries to accelerate their rate of industrialization, there were few such usable developments in Japanese technology.[7]

Of particular concern to the Division's staff was its observation that industry did not seem to recognize the importance of the research laboratory in industrial organization. Elsewhere in SCAP documents, however, staff members of the Division note that many parent companies felt that research should receive a low priority relative to immediate programs for increasing production. For example, many laboratories at the time were trying to synthesize sweetening agents or food products to relieve shortages.[8] During this period, some devices created during the war were being used to locate schools of fish and to develop uses of seaweed for food.

Successful experiments were also carried out on the utilization of solar energy to produce protein through the growth of algae.[9]

It was not expected that the SL Group would take up all the emerging issues. Rather, its goal was to create a forum that would help to bring issues to the fore. Kelly was hopeful that visiting U.S. scientists would help persons involved in the reorganization effort in Japan understand the importance of creating a linkage between research laboratories and industry.

After preliminary discussions, the SL Group developed a draft plan, which it submitted to GHQ in November 1946. The plan contained two primary components: to establish within the government a Science Agency staffed by capable scientists who would be charged with scientific and technical administration, and to create an advisory council to monitor the work of the Science Agency. The members of the advisory council were to be elected from the Japanese scientific community.

While the SL Group was developing its draft plan, two independent attempts to reorganize the national administration of science were underway. First, a resolution to reorganize Japanese science and to concentrate its direction and control in a cabinet-level committee was unanimously passed by the Japanese Diet on September 27, 1946.[10] This action had been promoted by an organization, formed in June 1946, called the Scientific and Technical Policy Comrades Association (Kagaku Gijutsu Seisaku Doshikai). This association was instrumental in establishing a parliamentary Science Club, made up of members of the Diet who were interested in the use of science to facilitate Japan's reemergence as a strong state.[11] According to Kelly, however, because the resolution was a surprise to the SL Group and the occupation authorities, the plan to establish a cabinet-level committee was disapproved by SCAP as premature in mid-December.[12]

More problematic for the occupation staff was the evolving plan put together by the Ministry of Education, the leading members of the Imperial Academy, and the Japan Society for the Promotion of Science. This proposal called for abolishing the National Research Council and transferring most of its functions to the Japanese Association for the Advancement of Science, which was to be under the authority of a revised Imperial Academy and JSPS. This plan came as a complete surprise to the members of the SL Group, who according to Kelly knew nothing about it. If any of the three existing institutions was to survive the reorganization, the scientists would surely have preferred the National Research Council, which many of them considered to be the most effective of the three.[13] The SL Group issued a 15-point critical review of the plan, noting in particular the undemocratic approach taken in its development and its apparent failure to recognize the "national and international significance of the reorganization opportunity."[14] The Imperial Academy responded with equal surprise to the rejection of its proposal, since its members had understood that reorganization plans were to be initiated and formulated by the Japanese themselves.

Kelly took a hard line on the issue. At a meeting with NRC and Imperial Academy representatives, he did not mince words: "Your activities have so far been rooted in the initiative of only a fraction of the Japanese scientific community. We need a plan fully representing the consensus of all the scientists of the whole country. You may maintain your present organization until the new system is established, but you must continue your work with what I said kept in mind." In an article written in 1949, Kelly said: "It is not known whether greater understanding was obtained from the reply, . . . but it was not certain that the Academy truly represented the Japanese scientists."[15]

These differences were ultimately negotiated on November 27, 1946, at a meeting of occupation officials with representatives of the Ministry of Education, the NRC, the JSPS, the SL Group, and the Imperial Academy. Out of this meeting came a proposal for a Preparatory Committee of approximately 50 people to determine the qualifications of electors to a penultimate planning committee for the "renewal" of Japanese science.[16]

The process for selecting representatives to the Preparatory Committee raised concerns among the advocates for change, including Kelly, because it was controlled by the "old guard." Recommendations for membership were initiated by the director of the Bureau of Scientific Education within the Ministry of Education. His recommendations for representation on the committee were submitted to the Imperial Academy, the NRC, and the JSPS for approval. The recommended scholars came from the three Japanese organizations and from key research laboratories and institutes in both the natural and cultural sciences, including law, literature, economics, science, engineering, agriculture, and medicine. The selected scholars, in turn, recommended people from the fields of public administration and private enterprise.

Despite early concerns that the Preparatory Committee would cave in to traditional ways, it did manage to break through certain barriers. Between its first meeting on January 17, 1947, and the following August, it met more than a dozen times to develop a process for selecting representatives to the penultimate planning committee. In keeping with SCAP's policy of leaving the planning to the Japanese, Kelly did not attend the meetings unless invited for a specific purpose.

The procedure the committee developed was an election by mail to be held between April and August 10. The electorate was composed of all scientific societies having 500 or more members, of which some 50 existed at the time.[17]

The election process resulted in 108 members' being named to the "Science Organization Renewal Committee." Each of the seven traditional university "faculties" (law, economics, literature, engineering, agriculture, science, and medicine) was represented by 15 members. In addition, three members were to represent the Society of the History of Science of Japan, the Democratic Scientists' Association, and the Association of Democratic Scientists.[18]

The latter association, known as Minka, had been founded before the war by Marxist scientists; however, its membership was open to scientists and engineers who wanted to promote democracy in scientific laboratories, no matter what their political persuasion.[19]

Despite the attempt to ensure a broader franchise, the outcome was somewhat disappointing. Tokyo remained at the center of the scientific establishment, with more than three-fourths of the members; the University of Tokyo alone held over half the memberships.[20] Private universities won only ten seats, even though slightly more than half of Japan's science and engineering graduates came from those institutions in 1946.[21] Nevertheless, as Kelly noted in a 1949 article, the Renewal Committee represented a significant break with tradition. The average age of its members was 50—young by Japanese standards. Moreover, many of the University of Tokyo candidates and even some members of the Preparatory Committee were defeated in the election. Only six of the elected members came from the Imperial Academy, and only about half from the large National Research Council.

While this was going on, Kelly had gotten Frank Jewett, president of the U.S. National Academy of Sciences, interested in organizing a committee of prestigious scientists and engineers to review the Japanese situation. Jewett had obtained funding from the Rockefeller Foundation for a six-week mission and had recruited six NAS members who were experienced in dealing with "people and problems at top level."[22] Roger Adams, head of the chemistry department at the University of Illinois, was asked to lead the group, which included W. V. Houston, president of Rice Institute; W. D. Coolidge, director emeritus of research at General Electric; W. J. Robbins, director of the New York Botanical Garden; Royal Sorenson, head of the department of electrical engineering at the California Institute of Technology and past president of the American Institute of Electrical Engineers; and Merrill Bennett, executive director of the Food Research Institute at Stanford University. Bennett, an economist representing the social sciences, wrote the group's final report. The inclusion of an "industrialist" (Coolidge) as well as an economist was considered by both Jewett and Kelly to be an important statement about "the need for R&D in the industrial rehabilitation of Japan."[23]

The group arrived in Tokyo on July 19, 1947, and stayed until August 28. Kelly had arranged an arduous schedule of nearly constant rail travel. Roger Adams described the first weeks of the trip in a report to Frank Jewett: "Following [the orientation meetings at GHQ-SCAP] a special train was provided with bedroom facilities and an excellent observation car to visit in succession Nagoya, Kyoto, Osaka, Hiroshima and Fukuoka. At each place we spent one or two days discussing with the university and industrial scientists the same problems that we did in Tokyo, namely, how science should be reorganized in order to satisfy the scientists of Japan."[24]

The group met with scientists at private and public research laboratories, institutes, and universities. It became quite clear that scientists throughout the country were eager for a change but not sure what that change should be. As Adams noted, "All in all our trip was most successful in that we crystallized in our minds the primary desires of the scientists away from Tokyo. In practically no case, however, had their plans been sufficiently formulated so that we could envisage a practical plan for reorganization."

The "Scientific Advisory Group," as the mission was called, concluded its six-week visit by attending the inauguration of the Renewal Committee. The members of the group played an important role not only by supporting the general direction and goals for reorganization that had been identified by the Liaison and Preparatory committees but also by adding their good will and prestige to the reorganization effort. Advocating a program of reorganizing science in its broadest sense, including the social as well as the natural sciences, the Scientific Advisory Group left its report with General MacArthur, who received it as "reference data helpful to the study and analysis carried out by the GHQ authorities concerned, the Japanese Government authorities and the Japanese Government institutions to assist in the reconstruction of scientific research as well as technique and attainment of an improved education system."[25] In the report, the SAG expressed concern about the strong feudalistic and bureaucratic tendencies it observed in Japanese culture and encouraged efforts to organize science and technology without those characteristics. The group clearly supported Kelly's view that the primary task before Japan's scientists was to help meet the immediate needs of the Japanese people.[26]

The Renewal Committee

The Science Organization Renewal Committee was convened by the Preparatory Committee on August 25, 1947, at the prime minister's official residence. In addition to Prime Minister Tetsu Katayama, the Japanese government was represented by Takeo Miki, the minister of communications. Renewal Committee member and future Nobel laureate Hideki Yukawa opened the meeting and introduced the prime minister. It was an auspicious moment, filled with hope.

Outside the walls of the prime minister's residence, the people of Japan were midway through the most dire of the postwar years. Despite severe hardship brought about by food and housing shortages, over 100,000 Tokyoites had turned out on a cold and rainy day in May for the ceremony to inaugurate the country's new constitution.

The occupation authorities were represented at the inaugural ceremonies by Harry Kelly and his superior, Brigadier O'Brien, who stressed the importance GHQ attached to the work of the Renewal Committee and the high expectations the authorities had for its work. The prestige of the committee was greatly enhanced by the presence of Prime Minister Katayama and the head of the NAS delegation, Roger Adams. In expressing his interest in creating a bridge between science and government, the prime minister lent his support to the realization of the partnership that Tamiya, Kaya, and Sagane had advocated since the beginning of their involvement with the Americans. He also advocated seeking strong public support for the deliberations. Many of the Japanese scientists attending the inauguration ceremonies told Adams that without the presence of the American scientists, the reorganization plans would not have come to the personal attention of the ministers.[27] In his remarks, Adams stressed the importance of R&D to the reemergence of Japan, emphasizing the crucial link between science and industry. He noted: "The dearth of foodstuffs and raw materials make imperative the exportation of manufactured goods in order to permit imports of necessary commodities. By science and invention arising from the cooperation of the scientists with industry, much can be done toward a balanced economy."[28]

At the conclusion of the ceremonies, the committee held its first meeting and elected its chairman, Kankuro Kaneshige. The result surprised virtually everyone, not least Kaneshige himself. A little-known 45-year-old professor of engineering at the University of Tokyo, Kaneshige had earned the respect of his colleagues on the Preparatory Committee through his consistent impartiality and attention to detail. Seiji Kaya was elected committee vice chairman for the natural sciences division.

Subsequent meetings of the Renewal Committee were held monthly in the Japan Academy's main auditorium. Funding for the plenary meetings (which would last two to three days at a time) and for the numerous subcommittee meetings was provided out of the prime minister's cabinet budget, and the administrative services were provided by the National Research Council. The occupation officials were asked to comment only on specific invitation of the committee.[29] Kelly was invited to address the second plenary meeting, and he stressed the need for an effective, versatile, nationwide science organization that could examine all aspects of Japan's scientific capability and address the country's pressing needs. As the SAG had done in its report, Kelly emphasized that not only the natural sciences but the entire realm of learning, including the social and cultural sciences, should be represented in the organization to be created by the committee.

In response to Kelly's address, the Renewal Committee drafted a formal written statement declaring: "With the support of Japanese scientists at large and of the GHQ, the Science Organization Renewal Committee is resolved to spend every effort in drawing up the charter for a new science organization that we may consider to be the best. And we are fully aware that this is our serious responsibility toward the Japanese people, and toward humanity."

Despite daily hardships and financial difficulties, the Renewal Committee met regularly for seven months. By the end of March 1948, a proposal and a report of the committee's activities were submitted by Chairman Kaneshige to the new prime minister, Ashida (who, like Katayama, headed a coalition government that included socialists). The proposal essentially affirmed the Science Liaison Group's recommendations for the establishment of a new Science Council and a governmental advisory body to be called the Scientific and Technical Administration Commission (STAC), but with some modifications. The SL Group had recommended a science agency staffed by capable scientists within the government, to be charged with scientific and technical administration, and an elected advisory council to monitor the work of the agency. This would have centralized the administration more than the Renewal Committee's proposal and would perhaps have transferred responsibility for the funding of research projects from the Ministry of Education to the new agency.

A bill to establish the Science Council of Japan was submitted to the Diet on June 30, 1948, and approved on July 5. The Science Council was to be a deliberative and independent consultative body. Its main function was to promote science and technology in Japan, but the law also provided that the government could request opinion and advice from the council on funding for scientific research and on any matter of scientific or technological importance deemed appropriate by the council itself.[30] The Ministry of Education controlled most of the government research funds through its oversight of the universities, but other ministries were also involved through their support of public institutes and laboratories. In addition, the Science Council was authorized to make recommendations to the Ministry of Education about the membership of the committee responsible for distributing grants-in-aid for scientific research.

Discussion of the governmental coordinating body recommended by the Renewal Committee was postponed until after the law establishing the Science Council went into effect. Even before the council elections, however, the government proposed the creation of a Scientific and Technical Administration Commission (STAC) in the prime minister's office to fulfill those functions. The law became effective on December 20, 1948.

In August the cabinet approved an allotment of 9.55 million yen from the Reserve Fund to cover expenses for the election. The Japanese National Research Council was dissolved, and its activities absorbed by the Japanese Society for the Promotion of Science. The JSPS became a private and independent body that could receive money from the government only for specific contracts. In practice, though, the JSPS remained a government instrument and was subsidized under a separate budget item, "Subsidies for Private Research Laboratories."[31] The Imperial Academy was made an honorary body within the Science Council.[32]

Organization of the Science Council

The Science Council of Japan Law formally established Japan's first democratically organized body authorized to advise the government on scientific matters of public importance. Reflecting widespread sentiment in favor of explicitly peaceful governmental pursuits, the preamble to the law outlined a series of lofty aims: "The Science Council of Japan shall hereby be established, on the conviction that science is the basis of a cultural nation, and having as its mission contribution, by joint will of scientists throughout Japan, to the peaceful rehabilitation of this country and promotion of the welfare of human society as well as contributing to progress in science by cooperating with the academic circles of the world."[33]

Once the law was passed, an elaborate election process began. This process started with district and national nominations. The law called for the election by peers of scientists from throughout Japan. Those elected would serve in one of seven divisions of the Council, organized under two broad departments: one for the cultural sciences and one for the natural sciences.[34] The Cultural Science Department included the division for literature, philosophy, and history, the division for law and politics, and the division for economics and commerce. The Natural Science Department had four divisions: one for fundamental science, one for engineering, one for agriculture, and one for medicine, dentistry, and pharmacy. Each division was entitled to 30 members, for a total of 210.

The Second U.S. Science Advisory Group

In the meantime, Kelly had made arrangements for a second Science Advisory Group to be appointed by the National Academy of Sciences. The three-week mission was headed by Detlev W. Bronk, Chairman of the National Research Council and president-elect of Johns Hopkins University. He was accompanied by I. I. Rabi (winner of the 1944 Nobel Prize in physics), Roger Adams (who had headed the first SAG), Zay Jeffries (a vice president of General Electric and general manager of that company's chemical division), and Elvin C. Stakman (chief of the division of plant pathology and botany at the University of Minnesota and president-elect of the American Association for the Advancement of Science). Their formal purpose was to review what had been achieved since the last NAS mission. The mission left for Tokyo on November 21, 1948, arriving just as approximately 40,000 Japanese scientists, representing all fields of research, were participating in the election of the Science Council.

From Kelly's point of view, the mission was a success in every way. The American scientists helped foster an awareness of the status of science in the United States and added to the general prestige of the new Science Council. They emphasized the important role science and technology would play in the reconstruction of Japan and suggested that the success of that role would depend in large measure on policies put in place by SCAP.[35] They also reported on how hard it was for Japanese students to get access to scientific equipment and literature and encouraged American assistance in the rebuilding of Japan's leading laboratories. Rabi, who pointed out that the United States had ample equipment but a shortage of scientific personnel, proposed that Japanese scientists be supported in U.S. universities for periods of two to three years at a time.[36] After the mission's inspection tours, Rabi also suggested that the United States "permit, encourage and aid the Japanese to establish one or more laboratories, particularly physics and communications" (modeled on the Telecommunications Research establishment in Great Britain during World War II) that could be used by American personnel in the event of a future military emergency in the Far East.[37]

Signs of Revitalization

As the mission headed back to the United States on December 20, 1948, the votes for the new members of the Science Council were being counted.

The members were elected for three-year terms through an elaborate mechanism of district and national nominations and voting. Of the 30 members in each division, seven were elected from district constituencies, one for each of the seven divisions. The remaining 23 members were elected on a national basis. For the first election of members, an Election Administration Committee, chaired by Professor Tadao Yanaibara of the University of Tokyo, sent out requests for nominations based on criteria specified in the new law to 1,750 learned societies and research institutes throughout Japan. The people whose names were submitted in response to the request were then sent voter registration cards. The lists of the "qualified" voters were publicly displayed in districts throughout Japan well before the December 20 elections. Nominations of candidates for office were filed by October 20, and 43,699 ballots were mailed out by November 12. Ballots were counted between December 20 and December 22, when the names of the 210 members of the Science Council were published.

The successful completion of the work of the Renewal Committee was the most obvious milestone in the postwar reorganization of Japanese science, but other events of 1948 also indicated signs of revitalization.

In May, Japanese scientists had organized large-scale expeditions to Rebun-To, a small northern Japanese island, to observe a solar eclipse. A significant aspect of the expedition was the cooperation between Japanese and American scientists in setting up the observation site.[38]

The Science and Technology Agency (initially called the Agency of Industrial Science and Technology) was established on August 1, 1948, within the Ministry of Commerce and Industry (reorganized as the Ministry of International Trade and Industry on May 25, 1949). Initially focusing on mining and manufacturing, the new agency was set up by the Diet to identify what assistance was needed to improve Japanese industrial research and production.[39] Japan's economic plan for recovery was now beginning to unfold, with an emphasis on improving heavy industry and international exports.[40] In this context, the Ministry of Commerce and Industry, and later MITI, took the lead among individual government agencies in implementing an emerging policy for scientific research that emphasized practical and immediately useful developments rather than purely academic basic research. The Ministry provided assistance to private industry to build pilot plants for new processes and to aid individual applied research projects.[41] At the same time, an Agency of Smaller Enterprises was created within the Ministry to help small entrepreneurial industries address financial, managerial, and technical challenges.[42]

This view emphasizing practical R&D over purely academic research was shared by many members of the newly created Scientific and Technical Administration Commission. Established within the Prime Minister's Office, the STAC was designed to act as a conduit for JSC opinion into the decision-making bodies of government. However, since STAC was composed of both elected scientists and appointed bureaucrats, sustained unanimity of opinion concerning priorities for the allocation of research resources was unlikely. STAC was broader in scope than the Renewal Committee's proposed coordinating body in that it was not limited to dealing with matters of interest to more than one ministry and in that it was specifically given an interest in international scientific matters.[43]

Thus, by the end of 1948, an institutional structure for the organization of science was in place, although authority for advising the government was dispersed among the Ministry of Education (which retained administrative control of government funds for scientific research in universities), the Ministry of Commerce and Industry (which operated 13 laboratories in various technical fields that were becoming increasingly important with respect to the economic recovery), and the newly created advisory and consultative bodies under the Prime Minister's Office.

Functionally, the Science Council appeared to be the most important advisory body for scientific matters; in practice, however, it was clear that the agencies representing the scientific interests of the ministries would also play an increasingly important role. By then the priorities of both SCAP and the Japanese government had shifted from dealing with political and social concerns to strengthening the economy and the standard of living. In that context, the ministries that were concerned with meeting the nation's immediate needs would command considerable influence over the national research agenda.

The Science Council was launched at a time of a subtle but perceptible shift within Japan's political environment and in the relationship between the United States and Japan. First, the return of Prime Minister Yoshida Shigeru to power in October 1948[44] marked a shift from a coalition government that included the socialists to a conservative one. Second, the shift in SCAP's priorities from political to economic reform began to take hold.[45] Third, as relations between the United States and the Soviet Union disintegrated, the need to treat Japan as an independent partner in the region became clearer. It was against this backdrop of political change that the Science Council optimistically prepared for its first session on January 20, 1949.

At the very first planning meeting of the Science Liaison Group, Harry Kelly had admonished the participants to "set aside personal interests and preferences of where the most important science advisory body should be." The Preparatory Committee and the Renewal Committee ultimately decided to protect the neutrality of the body by keeping it independent of government administration. Although the Science Council was under the jurisdiction of the prime minister and financed out of the national budget, it differed from other administrative bodies. Since its members were not appointed by any government authority, they were responsible only to the scientists who elected them, rather than being loyal to any single agency or ministry. This helped to ensure the Science Council's objectivity, though the fact that it was anchored outside the Japanese power structure may also have contributed to weakening its influence in the long term.

A Japanese Framework for Science Policy

4

A Fresh Start

The Science Council of Japan and the Scientific and Technical Administration Commission were entirely novel entities in Japanese history. A great deal of excitement surrounded the inauguration of the Science Council because it represented a unique approach to the government-science relationship in Japan—an approach that offered the possibility of coordination and the hope that decisions about government-funded research would now be made by active first-rate scientists. The enthusiasm that marked the opening ceremony was captured in the *Quarter Century Annals of the Science Council of Japan:*

On this January 20, 1949, the members of the Science Council elected late in the previous year thronged to the Japan Academy auditorium in Ueno Park. It was a fine chilly morning.

In attendance on this occasion were 200 of the 210 members, evidencing the keen intent of Japanese scientists to set forth on a new evolution of scientific research in Japan, at this turning point of history brought by our defeat in the war.

> *The assembled members not only placed their hopes on this newly organized system, but were keenly conscious of creating a system that was unprecedented, and the proceedings were set forth from the outset in a completely new pattern.*

The first business meeting began with the election of a president and two vice presidents (one for each major division). After lengthy debate, Naoto Kameyama was elected president. Yoshio Nishina and Sakae Wagatsuma were elected vice presidents for the natural and cultural sciences, respectively. After the plenary session, the members broke into their division units to elect officers. Seiji Kaya was elected president of the Division of Natural Sciences.

When the elections were over, Kaneshige informed the full assembly of the elaborate procedures that had led to the formation of the Science Council. The road had not been smooth. Rivalries and tensions over control of the Council had had to be negotiated. As mediator among the various positions, the Scientific and Technical Division played an important although indirect role in determining the outcome. From the formation of the Scientific Liaison Group through the work of the Preparatory and Renewal Committees, the not-so-invisible hand of the American authorities in SCAP is apparent in the legitimation of some groups as opposed to others, in the selection of advisers, and in the power of veto the Americans could exercise in advancing their own agenda for reform.[1] This is not to say that the Science Council and its membership were imposed on the Japanese. Quite the contrary. Kelly and his colleagues in SCAP as well as the visiting scientists who provided assistance to the Renewal Committee and the Science Council understood that if reform was to be effective and enduring it would have to come from the Japanese themselves and serve their interests. What developed in the reorganization process was a convergence of interests between moderates in the Japanese scientific community and moderates in the SCAP bureaucracy. These moderates shared the belief that Japan's economic future depended upon linking the policies intended to strengthen its capabilities in science and technology with its economic goals.

Despite the enthusiasm about new democratic procedures and the high hopes of many members that they could create a bridge of understanding between science and government, a thin veil of uncertainty hung over the proceedings. At the opening meetings, Kaneshige took pains to reassure the assembly of the Japanese origins of the Science Council while expressing appreciation for the role of the occupation authorities (particularly Harry Kelly) in bringing the council to fruition: "We all know that in order to make headway on any matter today in Japan, we require on many occasions the understanding and assistance of the General Headquarters of the Allied Powers, and the question of renewing the organization for science was no exception. In the present situation, as far as I know, and I think I probably know everything that concerns the question of renewing our organization for science, the part taken by GHQ was all assistance, and nothing but assistance."

Although Kelly did not take credit for having played a significant role in the formation of the Science Council, Kaneshige made it clear that he had been instrumental in securing the confidence of the occupation authorities in the plan proposed by the Science Liaison Group and the Renewal Committee—a confidence that was clearly necessary in order for any proposed changes to make headway while the country was occupied. In publicly thanking Kelly for his support, Kaneshige noted: "Dr. Kelly, Deputy Chief of the Scientific and Technical Division, never on any occasion stated what might be his own plan. Even when he was not quite of the same mind with what we proposed, he finally ended by respecting our opinion, and even took pains to bring the view of GHQ round to obtain their agreement. I owe whatever I have been able to accomplish in my task entirely to the fact that Dr. Kelly was at the post he occupies. If he were not there, I personally cannot help seriously doubting whether the Science

Council of Japan would have seen the light of day in its present form."

Inauguration

The official inaugural ceremony was held in the auditorium of the Science Museum, adjacent to the Japan Academy. In his last function as chairman of the Renewal Committee, Kaneshige introduced the new Science Council's president, Naoto Kameyama. In keeping with the lofty aims of the council, Kameyama expressed the desire that world peace be maintained, that justice be upheld, and that, in light of the austere realities facing Japan, science should serve as the basis for rehabilitating the country and contributing to peace and prosperity. Prime Minister Shigeru Yoshida was represented at the ceremony by State Minister Shunkich Ueda, and the Diet was represented by the Speaker of the House of Councilors, Tsuneo Matsudaira.

The General Headquarters of the Occupation was represented by Harry Kelly, who commended the Renewal Committee not only for developing the tools necessary for attacking technical problems but also for the example the members had set in reaching consensus despite formidable challenges and adverse circumstances. Remarking on the aims of the Science Council, set forth in the preamble of the law in which it was created, Kelly said: "These are bold words, and it will take strong men to fulfill the obligations assumed. In fulfilling these responsibilities, the individual members of the Council must first transcend their particular scientific fields of interest and locality."

Kelly went on to reiterate his own major objectives and SCAP's emphasis on economic development: "It is well recognized that science and technology are essential to the economic recovery of Japan. The new Council must recognize the technical problems indigenous to the present conditions of Japan. The solution [to] these problems will be [advanced] only by developing a cooperative spirit among scientists, industrial concerns, and the government. This industrial application, in addition to being economically necessary, will also be helpful in further developing the prestige of the fundamental scientists and developing conditions necessary for their continued progress."[2] The alliance to which Kelly referred had already been highly successful in the United States, where government, science, and industry had jointly produced the advanced technologies that were instrumental in winning the war and in establishing the nation's scientific superiority.

In emphasizing the importance of "industrial application," Kelly was also indirectly addressing a controversy that had dogged the Renewal Committee's meetings: whether or not to include engineering in the Science Council's definition of what constituted "science." A particularly heated argument had involved Kelly's good friend Sagane and Sagane's ally Eizaburo Nishibori, a former Kyoto University science faculty professor who was then doing research on vacuum tubes for the Tokyo Electric Company (today known as Toshiba). Sagane and Nishibori advocated including applied engineering in the Council's definition of science. At the time, Nishibori headed an organization that brought together scientists and engineers from diverse fields to promote quality-control practices within private enterprise.[3] He saw in the formation of the Science Council an opportunity to advance the interests of the applied side of R&D through the Council's potential influence on government policy. But when he and Sagane proposed a Council group on technological development, their views were received with derision. "The assembled gentlemen of learning called what I described 'trash science,'" Nishibori later recalled. Only the intervention of Professor Otaka, who headed the social science division of the Renewal Committee, quelled the uproar. Pointing out that the two Japanese ideograms for the word *gakujutsu* (science) denote learning and technique, Otaka proposed that the Council embrace both concepts —science and engineering—in its composition. Nishibori notes in his memoirs, however, that in the end "even the Engineering Division members largely represented the 'learning' rather than the 'technique' (or applied) aspect." He continues: "Sagane and I were branded by these learned scholars of technology as radicals, and were shooed out of their company. For our part, we . . . have boycotted the Science Council ever since. Sagane soon departed for his second stay abroad, while I retired to the obscurity of Toshiba."[4]

The argument between Nishibori and Sagane and their colleagues on the Science Council illuminates an issue that was central to the struggle for control of the reorganization of science in Japan. "Science," for purposes of the structure of the Council, was interpreted broadly to include not only the natural sciences but also the applied sciences of engineering, agronomy, and medicine and the humanities, literature, the social sciences, and political science. But this inclusivity did not diminish core differences of opinion on the conduct of research: Should it be left to the creativity of individual scientists, or directed toward the attainment of social goals? Not all scientists who sought reform agreed with Kelly's prescription of scientific research to cure the problems confronting Japan. In 1949, only the fourth year since the war ended, some saw disquieting conceptual similarities between this approach and the wartime scientific mobilization effort. A leading proponent of this view was Hantaro Nagaoka, a pioneer of Japanese physics and Sagane's father.[5] The proposal may have raised the sensitive question of the link between technological development and militarism in the minds of some of the scientists at the Science Council's first meeting; others may have reacted out of conservatism in the face of a proposal for radical change. The reformers had made a decision to cast off the old academic ways espoused by Nagaoka and others. The concept of change is central to the scientific enterprise, but change is necessarily fraught with uncertainty. As Japanese science took the first major step in its transformation from old-style academic traditions to postwar modernism, the Nishibori-Sagane proposal may simply have been too much too soon. No subdivision devoted to technological development *per se* was created. (The subdivisions of engineering that were eventually created included applied physics, mechanical engineering, electric engineering, naval architecture, civil engineering, architecture, mining, metallurgy, and applied chemistry.)

From Proposal to Reality

It was clear from arguments like the one over technological development that the lofty aims and inclusive structure of the Science Council did not resolve the conflicts inherent in implementing the goals and objectives of the newly created entity. Heated debate continued for three days. Was the equal allotment of members to each of the seven divisions appropriate, or should the council consider proportional representation? The engineering division, for example, represented more than 24 times as many voters as the law and politics division, yet each profession was represented by the same number of members on the council. Was the classification of divisions itself appropriate? The seven divisions reflected the academic structure of the seven faculties of the University of Tokyo. Did this strengthen or weaken the influence of the council? Where did the council want its influence felt most, and how would this be accomplished? Addressing these and other questions, the members spent long hours in uncomfortable circumstances. The meetings were held in a frigid auditorium. Breaks in the deliberations found the members huddled around temporary charcoal heaters set up in the halls of the academy. Dressed in overcoats, they painstakingly discussed every motion put before them outlining the functions and procedures of the Science Council. Perhaps as a hedge against the cold, the members seemed to increase their smoking. The cigarette smoke grew so thick that at the second day's meeting Kaneshige, himself a heavy smoker, proposed that smoking be banned during plenary sessions. The rule, adopted unanimously, is still in effect. The *Quarter Century Annals* record that "despite such austerity, the Council members assiduously attended to the proceedings through the three days of session, loath to neglect any detail. Their heated deliberations can be retraced today in the official minutes of the proceedings, but strangely, no photograph remains of the assembled members. No one seems to have thought of the assembly as a festive event to be remembered with a souvenir picture."

The Functions of the Science Council

The Science Council of Japan was set up to serve as the official domestic and international voice of Japanese scientists. As an advisory body, the council was to focus on the distribution of government grants and subsidies for research, on the administration and the budgets of governmental institutions and laboratories, and on providing advice on the education and training of scientists in universities and research institutes.[6] The law, however, did not *require* the government to seek such advice. One review of the draft versions of the bill found in the SCAP files of the Scientific and Technical Division suggests that an early version was worded to the effect that the government "*shall* seek the opinions of the Science Council of Japan on the following" and that this was later changed to "*may* seek the opinions."[7] The draft change to nonbinding authority became policy when the law was passed in its revised form.[8]

In addition to the nonbinding authority, the Science Council would also face the challenge of maintaining its autonomy despite government funding. The Renewal Committee had expressed its preference that the Council be financially independent of the government, but under the current economic conditions this was impossible.

The Science Council took up its tasks immediately. In addition to advising the government through the office of the prime minister, the council was expected to promote international cooperation and coordination of scientific research. As a first step, it became a member of the International Council of Scientific Unions.

Harry Kelly's office played an important role in helping the Science Council obtain international recognition. As the Council was meeting for the first time, plans were being made by the Scientific and Technical Division to deliver papers prepared by Japanese scientists at the Seventh Pacific Science Congress, to be held in February 1949. This important regional meeting was sponsored in part by the Far East Commission. Among the issues to be addressed were plans and programs for the development of research in the Pacific area in general and for marine science in particular. Although Japanese scientists had much to offer at the conference, the other countries were reluctant to include Japan in the deliberations. The External Affairs office of the host country, New Zealand, said it was "inappropriate that Japan should attend" because a peace treaty had not yet been signed.[9] After much negotiation between SCAP and the conference organizers, it was decided that SCAP would represent the Japanese scientists at the conference. Abstracts of their research were prepared and transmitted to the conference participants by SCAP officials from Kelly's office. In the end, more than 140 papers were delivered to the Pacific Science Congress by GHQ scientists.[10] Paul Henshaw, a biologist on the staff of the Scientific and Technical Division, attended as one of four SCAP observers and was instrumental in getting the Science Council of Japan recognized by the Pacific Science Association.

The Science Council seemed to fulfill its expectations. As it gained international recognition, it provided an institutional basis for foreign travel by Japanese scientists. Within Japan, in the first two years of its existence, it received 22 inquiries from government and issued 75 spontaneous recommendations and opinions on science-related issues. The committees dealing with public administration of science and with nuclear energy were particularly active in the early years.

Ironically, SCAP seems to have deemed the Science Council a success for having influenced Japan's pursuit of a technology policy that reflected the very approach Sagane and Nishibori had proposed. A draft report of the Scientific and Technical Division for the 1949 Annual ESS Historical Report noted: "The course of governmental activities has been influenced to a significant degree by JSC. It has been effectively fostering a trend toward a greater diffusion of the results of scientific research into the technology of industry."[11] The draft report also praised the Science Council for "helping to bring about the acceptance of the principle that importation of technology and travel of engineers and scientists abroad should have a high priority in the allocation of Japanese foreign exchange. . . . JSC's relation with the government has been especially active in such matters as national financing of research work, administrative treatment of research workers and reorganization of government establishments bearing on research work."

The first annual report of the Science Council highlights similar achievements and goals, noting the establishment of a committee for "promotion of industrialization of research results." The problem the committee was charged to address was the long-standing need to reduce the amount of time between a product's development in the laboratory and its commercialization. The Science Council's committee to promote the industrialization of research results found that the most important reason for industry's weak research performance was the lack of funds necessary to test potential products for commercialization. As a first step toward solving this problem, the Science Council supported the creation of a "Development Bank for Industrial Technology."[12] The Japan Development Bank was created in 1951 and became, in the words of one Japan scholar, "one of MITI's most important instruments of industrial policy."[13]

It should be noted that the Science Council itself did not single out its work and recommendations on industrial research from the other accomplishments of its first year. In fact, it paid more attention to its organization and to its efforts to promote academic freedom and obtain increases in the research budgets of the major funding ministries: Education, MITI, and Agriculture and Forestry. The report also notes the resolution, adopted at the sixth meeting of the Science Council in April 1950, to refrain from any scientific research that contributed to war: "At the inauguration in January 1949, the Science Council of Japan declared within the country and abroad its firm conviction that science should build the basis of an enlightened country and peace of the world. Out of our heartfelt wish to avoid the misery of war, we . . . express our firm determination not only for realizing our former statement, but also . . . that we would never be engaged hereafter in scientific researches aimed at war."[14]

In the first two years of its existence, the Science Council received, on average, one request per month for advice to the government, and it offered 71 analyses of government issues between January 1949 and the middle of 1951.[15] But in the ensuing years, its influence waned quickly. Inquiries from the government between 1951 and 1955 averaged less than three per year, and after that only one or two per year. By the 1960s, the need for the Council was being questioned publicly. It survives today primarily as a vestige.

What caused this decline? Several reasons have been advanced, including the weak authority the Science Council was granted in the final legislation and its relative isolation from the Japanese government after GHQ-SCAP rejected the Diet-approved plan under which science and technology would be controlled by a cabinet ministry.[16] Anchored on the periphery of Japan's emerging institutional elites (the bureaucracy and the emerging conservative coalition), the Science Council retained its autonomy but lost its influence.

The politicization of the membership through the election system may have been another weakness. Designed to promote freer participation by active scientists, the three-stage election system of nationwide, district, and specialized sectors provided abundant opportunities for politicking and favoritism. Beyond that, most of the elected members were politically to the left of the conservatives, who had regained political power with the second Yoshida government in October 1948. According to one SCAP report, the Communist Party took an active part in the first Science Council election by supporting 61 candidates in 26 of the 43 fields of specialization, of whom 26 were elected. The report goes on to note that "in addition, there were other candidates known to be Communists or considered as strong Communist sympathizers, whose election brought the total Communist strength to an estimated 40 with representation in each of the seven divisions of the Council."[17] Harry Kelly was concerned about the membership elections and the fate of the Council after he left Japan in 1950, and always inquired about it when he returned. He shared the disappointment of the founders. Echoing Tamiya's early criticisms of Kelly's idealism, he once exclaimed to Tamiya: "What idealists we both were at that time!"

Another reason for the Science Council's decline in influence may be rooted in bureaucratic competition. With emphasis increasingly placed on industrial rather than academic research, government ministries other than the Ministry of Education became interested in the research budget and sought to control its disbursement. Anticipating rivalries and complications between the ministries, the Renewal Committee had recommended establishing a body to coordinate their scientific interests. This recommendation resulted in creation of the Scientific and Technical Administration Commission (STAC) in March 1949. Although representatives of 13 ministries made up half of STAC's membership, government authorities were not enthusiastic about the new agency.[18] The first annual report of the Science Council ends on this dubious note: "It was no wonder that the administrative authorities were . . . not sympathetic to STAC. However, their understanding as to the true nature of STAC as well as their recognition of the effectiveness of its work have been gradually increased. Still, it will take several years perhaps before STAC will be able to display its full power. It is sincerely desired that STAC can eventually contribute to the rehabilitation of Japan as it was expected and projected by scientists."[19]

The Scientific and Technical Administration Commission

STAC (as it came to be called by both Japanese and occupation personnel) was created both as a conduit for the views of the Science Council and as a vehicle for the coordination of science- and technology-related activities within the Japanese government. The final report of the Renewal Committee identified the problems the agency was designed to correct:

In the adoption and implementation of administrative measures, there has been in the past a certain deficiency in making expeditious and effective use of the results available from scientific research. What is more, the administration of scientific and technical matters is not properly coordinated between the different branches of government, which results in a dissociation between science and national policy, and in a general absence of the scientific approach to administration.

Promotion of the different domains of basic science also has missed receiving due recognition and active support by government; the consequent scanty equipment and the meager funds and supplies made available for research have clipped the wings of research and undermined the ardor of scientists for study.[20]

Initially successful (along with the Science Council) in increasing research budgets, STAC soon saw its influence wane like that of the council. Set up as a small agency within the prime minister's office, STAC was no match for the expanding bureaucracies of the ministries. It got off to a rocky start because of misunderstandings concerning the overlap of its mission with that of the Ministry of Education, which had traditionally controlled government-sponsored research.[21] This tension was felt perhaps most keenly by Kaya, who, while retaining his professorship at the University of Tokyo, was also serving as director of the Scientific Education Bureau of the Ministry of Education during the period when the Science Council was launched. Well aware of the lack of coordination and the limited government funding for research, Kaya had engaged Kelly in his efforts to lobby the Yoshida government for an increase in the science education budget. Kaya arranged for Kelly to join him and Nishina at a luncheon meeting with the prime minister. Kelly tried to engage the prime minister in a discussion about promoting scientific research, but Yoshida adroitly avoided the topic.

Another attempt to gain government support for science during Kaya's tenure at the Ministry of Education appears to have met with greater success. Together with Hitoshi Kihara, a prominent Japanese geneticist, Kaya and other scientists approached the prime minister with the idea of establishing a national institute for genetics. Genetics had been an issue of significant political dimensions since before the war, and certainly was in its aftermath,[22] but on this occasion the scientists wanted to press the potential agricultural benefits of genetic research. To dramatize this potential, they brought along a seedless watermelon. The story has it that Kihara swore to the prime minister that if the melon contained any seeds he would commit hara-kiri with the rind. Fortunately for Kihara and the institute's proponents, the melon was seedless and the prime minister was duly impressed.

Agreement was reached to establish the genetics institute at Mishima in Shizuoka Prefecture, but one hurdle remained: the GHQ's approval of the project was required. Unfortunately, Kelly was in the United States at the time, and other officers who were unfamiliar with the project and did not personally know Kaya refused authorization on the grounds that Mishima was not the right location. It was not until a few days later, when Kelly returned, that the issue was resolved and authorization granted—a move that was in keeping with the Scientific and Technical Division's policy of leaving such decisions in the hands of the Japanese. The decision also reflected the mutual respect Kelly and Kaya had for each other—a respect that had grown as they worked together, facing many thorny issues, over a three-year period during which each of them had to find ways to generate enthusiasm and foster a spirit of cooperation among his colleagues.

Kaya described the nine months from September 1948 through the following May as "the thorniest I ever lived through." Despite the resentment of his colleagues at the Ministry toward the newly formed Science Council, Kaya managed to establish a working relationship between the two entities—a relationship that entailed formal inquiries for advice from the Ministry of Education to the Science Council on such sensitive issues as grants-in-aid for scientific research (previously the sole responsibility of the ministry) and a pending draft "Law Governing the Administration of Universities." Rather than avoid the conflict that the new relationship imposed on him, Kaya seems to have confronted it with considerable diplomatic skill. This skill would serve him well in facilitating future Science Council initiatives, including the increased provision of support for nuclear research and the launching of Project Antarctica.[23]

As far as the occupation authorities were concerned, STAC's most effective work during the first year was its effort to "gain acceptance in government circles of the principle that emphasis in the scientific field must be shifted from academic research to practical and more immediately useful research and development."[24] The 1949 report on the progress of the Science and Technical Division noted "evidence that this principle has been accepted by all ministries including the Ministry of Education which has now acceded despite the fact that by doing so it loses some of its traditional control over the research activities of the nation."[25]

In keeping with Japan's policy of emphasizing international contacts to enhance its export markets, STAC became actively engaged in identifying international opportunities for Japanese scientists, including travel abroad and the importation of technology and research materials. The Scientific and Technical Division actively supported these efforts, securing funds to hire expert technical advisors for key Japanese industries and to "provide the means whereby Japanese industrial engineers could observe American industrial techniques which may be considered adaptable to Japan's industrial practices."[26] The Division was also able to support Japan's participation in international congresses, payment of royalties on technical service and patent license agreements, and support for applications for foreign industrial property rights.[27]

STAC was primarily a consultative body. Its 26 members were equally divided between government representatives and scientists nominated by the Science Council. The prime minister served as chairman, and a cabinet minister as vice chairman. Monthly meetings were held to forward recommendations from the Science Council to appropriate government agencies. A dozen subcommittees were set up to deal with specific issues, including the preparation of draft proposals for the government's science and technology budget, the disposition of inventions made by public servants in the course of their professional duties, and, most important, the importation of foreign technology, scientific materials, and radioisotopes.

Many of these STAC functions overlapped with those of the newly created Industrial Technology Agency in the Ministry of Commerce and Industry. At the end of the occupation, the government proposed the formation of a Science and Technology Agency to deal with industrial development as well as with the comprehensive planning of government policies in science and technology. The proposal was debated for several years within the

Science Council. Many academic scientists, whose intention in creating the Council on the periphery of official administration was to buffer it from politics, expressed concern about the negative potential of strong political control over science and technology. Nonetheless, in 1956 a law was passed by the Diet, and the Science and Technology Agency was launched to promote the development of science and technology with a view to improving the national economy. Finally, in 1959, a Council for Science and Technology was created in the prime minister's office as a overarching consultative body. The membership of this council includes the heads of the Japan Science Council and the Science and Technology Agency and three independent experts.

A piece of Nishina's cyclotron being dumped into the Bay of Tokyo, November 1945. (North Carolina State University Archives)

A photo from the trip to Hokkaido, February–March 1946. Among those pictured are Yoshihiro Asami (third from left, seated), Harry Kelly, and Takashi Minoshima (far right, seated). (North Carolina State University Archives)

The first Scientific Advisory Group at Mikimoto's Pearl Culture Station, July 31, 1947. Left to right: Royal Sorenson, Merrill Bennett, W. V. Houston, Mikimoto, Roger Adams, W. J. Robbins, and W. D. Coolidge. (U.S. Army Signal Corps)

A caricature of Kelly by Horishi Tamiya, 1947.

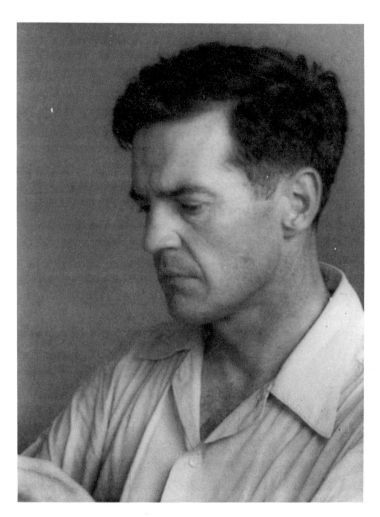
An informal portrait, Tokyo, 1948.

The second Scientific Advisory Group at Hokkaido, 1948. Detlev Bronk is at left; Zay Jeffries is fifth from left; I. I. Rabi is seventh, Roger Adams and E. C. Stakman are ninth and tenth. Harry Kelly is at right.

Kelly, Nishina, and Rabi during the visit of the Second Science Advisory Group, 1948.

At the first meeting of the Science Council of Japan, January 21, 1949. Kelly is congratulating Naoto Kameyama, who has just been inaugurated as president. Others (left to right): Kaneshige, Wagasuma, and Nishina. (U.S. Army)

Kelly visiting the family of Horishi Tamiya, summer 1949. (Kelly family album)

Kelly surrounded by Japanese reporters on his arrival for the third meeting of the U.S.-Japan Committee on Scientific Cooperation, 1963.

Kelly introducing Kankuro Kaneshige, co-chairman of the U.S.-Japan Committee on Scientific Cooperation, to U.S. Secretary of State Dean Rusk, July 9, 1968. (U.S.Department of State)

The memorial stones of Harry C. Kelly (left) and Yoshio Nishina (right) at Tama.

Kelly and Nishina

5

Yoshio Nishina met Harry Kelly in the aftermath of the cyclotron incident, soon after Kelly and Fox had arrived in Tokyo. Nishibori introduced them, and soon Nishina was taking an active role in both the Science Liaison Group and the Science Organization Renewal Committee. In a 1975 interview, Kelly recalled the affection he felt for Nishina and other scientists with whom he shared the intense occupation experience. "When I would talk with anybody like Nishina, I would trust him like a brother," Kelly said, adding "or Kaya, Kaneshige, Sagane."[1] Nishina and Kelly shared a humanistic vision of the scientific endeavor that helped them develop a strong personal relationship based on mutual understanding and respect. In addition to their work on reorganizing science, the two men proved to be effective in helping to provide the materials Japanese scientists needed for their research.

Even before the reorganization initiatives were underway, Nishina impressed upon Kelly the urgency of more immediate and pressing concerns. One was the need for radioisotopes for research in agriculture, medicine, and other areas. The other concerned the survival of Riken (the Physical and Chemical Research Institute in Tokyo) in the wake of the early occupation policy of liquidating the *zaibatsu* (conglomerates).

Radioisotopes

Radioactive isotopes (radioisotopes) are essential to effective research in many areas of natural science and engineering. Before the advent of nuclear reactors, the most effective technology for producing artificial radioisotopes was the cyclotron, invented by Ernest Lawrence of the University of California at Berkeley. Lawrence had encouraged scientists from around the world to come to Berkeley, and Ryokichi Sagane had worked with him for a full year during the 1930s. Upon his return to Tokyo, Sagane was instrumental in helping Nishina build cyclotrons. Now, however, with their cyclotrons dismantled, Japanese nuclear physicists and other scientists could not produce the radioactive tracers they needed for their work.

Nishina's only hope for resuming the research that was underway when the cyclotrons were destroyed was to import radioisotopes from the United States. He submitted an official request to GHQ, and he repeated his request to Kelly and Fox every time he went to see them. In April 1946, SCAP granted him permission to prepare and use stable isotopes of such elements as carbon, nitrogen, and sulfur,[2] but distribution of radioactive isotopes was not allowed until the following year. The U.S. Atomic Energy Act, passed in August 1947, provided for foreign distribution of radioisotopes, and the next month, at the Fourth International Conference on Cancer Research, President Harry Truman announced that the U.S. Atomic Energy Commission would accept requests from foreign countries for radioisotope allocation.

When the Scientific Advisory Group of the National Academy of Sciences visited Japanese laboratories in 1947, Nishina repeated his request and was reassured by the response he received. But nothing happened. The issue remained dormant until the spring of 1948, when Kelly's mission took him back to the United States briefly. Raising the question directly with authorities at the Atomic

Energy Commission, Kelly got a favorable reply. He returned to Japan hopeful that the details for the delivery of a limited supply of radioisotopes would soon be worked out.

Still, hurdles remained—notably how to pay for the materials. Kelly again stepped in with his usual personal business style, this time calling upon his friend William J. Robbins, director of the New York Botanical Gardens. Robbins, who as a member of the first National Academy of Sciences Scientific Advisory Group to visit Japan had been impressed by the intense work going into the renewal of Japanese science, suggested that a grant-in-aid be sought from the NAS, and he enclosed an application with his letter to Kelly. The application filed by Nishina was approved, and Kelly immediately requested the supply of radioisotopes. Scientific and Technical Division reports indicate that months of negotiations over delivery and payment ensued before final approval was granted in November 1949.[3] The political implications of the approval for both countries were not insignificant. Japanese press accounts noted that Japan was the first occupied country to be authorized allocation of the valuable elements. Public announcements in the United States noted that such shipments would be under constant supervision of the GHQ staff. Accounts also noted that the State Department had informed the Atomic Energy Commission that the sale was compatible with the general principles of U.S. policy toward occupied areas. The policy was to provide assistance that contributed toward the establishment of stable domestic governments.

Anticipating the benefits to be derived from the use of radioisotopes, the Science and Technology Division's 1949 progress report noted that "the scientific techniques made possible through the use of these materials are so outstanding that numerous experiments in applied science not otherwise possible will now be within the range of Japanese scientists and engineers. Such research may very well have a noticeable impact upon the economy of Japan within the next year or two."[4] When compared with the Economic and Scientific Section's response to Nishina's request to produce radioisotopes, this statement shows how far SCAP's policy on scientific matters had evolved by 1949.

The first shipment, which arrived on April 10, 1950, contained 6.2 grams (0.4 millicuries) of antimony-125 donated by the NAS to Dr. Nishina. The 40-kilogram cargo was handed over to three researchers from Riken, who shouldered it between them on a *tembin-bo* (carrying pole) before transferring it to a tricycle and then to a train for transport from Yokohama to Tokyo. At the University of Tokyo, the packing crate was opened with great ceremony. The newspaper *Asahi Shimbun* reported the story that day under the headline "Vanguard of Atoms for Peace Mission—Radioactive Isotopes Long Awaited by Our Scientists."

Although the isotopes were received ceremoniously, handling them was another matter. At that time in Japan, there was no readily available radioisotope cask. A creative member of the lab staff was able to improvise one from some old lead he had found in a junk shop. He melted the pipe to fabricate a lidded bowl, into which he gingerly placed the radioisotope.

Further shipments of carbon, phosphorus, sulfur, and iron isotopes were distributed to five different institutes for research in chemistry, biology, and agriculture. In addition to the NAS grants, $4,000 was set aside in the GHQ's commercial fund for additional purchases, and this led to regular importation of radioisotopes through commercial channels. Shipments were repeatedly reported in the press as "Missions of Atoms for Peace," emphasizing to the public the nation's commitment to exploring the constructive potential of atomic technology.

A committee headed by Nishina was set up within the Scientific and Technical Administration Commission to monitor distribution of the radioisotopes according to criteria set forth by GHQ. Those criteria listed economic rehabilitation, medical research, and clinical therapy as the only purposes for which the elements could be imported. Importation was on a government-to-government basis, with the materials forwarded directly from the U.S. Atomic

Energy Commission to GHQ in Tokyo, which then transferred the consignments to STAC. In addition to screening applications that were submitted through various government ministries, the committee reviewed progress reports on the use of the radioisotopes for research; the reports were then forwarded to the AEC.

The government-to-government approach continued until after the signing of the San Francisco Peace Treaty in the fall of 1951 and the departure of the occupation forces. Expecting that a significant increase in the demand for radioisotopes would accompany the transformation of importation from a government undertaking to an essentially commercial enterprise, the STAC began preparations for a new coordinating body. Kaya, who was then president of the University of Tokyo, became chairman of the coordinating committee after Nishina's death in 1951. Under his leadership a Japan Radioisotope Trade Association was created for researchers and technicians. The objectives of the association, which Kaya headed, were to keep the costs of radioisotopes reasonable and to develop and disseminate information about their hazards. As it grew, it began to organize seminars and issue periodicals as well.

Among the problems the new trade association encountered was the difficulty of collecting payments from impoverished university labs. Staff members often made the rounds of scientific institutes to collect. Sometimes the payment they received did not even cover their travel expenses. Kaya recalled the following story from those times:

Payments for radioisotope supplies lagged, and the University of Tokyo was the worst in this respect. So we decided to stop selling to the University; we drafted a letter from S. Kaya, President of the Japan Radioisotope Association, addressed to S. Kaya, President of the University of Tokyo [demanding payment]. I showed this draft letter—which was ready to be mailed—to the head of the University administration. He begged me to wait just a little while before mailing it. In fact, we did not need to send the letter, for the payments were made very soon after that.

In 1952 the Japan Radioisotope Trade Association moved its headquarters from the STAC office to the Nishina Laboratory at Riken. Kaya continued to serve as president until his death in 1988.

The importation of isotopes in the postwar years was an important factor in the revitalization of Japanese science because it stimulated research in new areas of medicine, agriculture, and technological development. It opened possibilities for nuclear research and became the cornerstone of modern Japan's nuclear science and technology.

Saving Riken[5]

The Physical and Chemical Research Institute (Riken) was an exception to the feudalistic, elitist tendencies that appeared to characterize Japan's scientific research establishment at the time of the occupation. Riken had been specifically set up to lessen Japan's dependence on foreign science and to spur industrial development.

Although such an institute had been proposed early in the century by prominent Japanese scientists, the need for it was not widely recognized among business and government leaders until World War I, when imports of military supplies, pharmaceutical products, industrial materials, and machinery virtually ceased.[6] A special committee was formed in 1914 to study the possible development of a chemical industry. This led to a proposal for an institute that would be financed jointly from private contributions and government subsidies. A statement of the research objectives of the proposed institute noted: "While the Institute for Physical and Chemical Research will carry on scientific research, it will also endeavor to respond to the needs of the state and society, aiming to promote public welfare."[7]

In 1917 the committee received permission from the Ministry of Agriculture and Commerce to set up the institute as a private foundation emphasizing the centrality of scientific research to industrial development. Noting Japan's scarcity of natural resources, the institute charter stated:

To promote the development of industry, the Institute for Physical and Chemical Research shall perform pure research in the fields of physics and chemistry as well as research relating to its various possible applications. No industry, whether engaged in manufacturing or agriculture, can develop properly if it lacks a sound basis in physics and chemistry. In our country especially, given its dense population and paucity of industrial raw materials, science is really the only means by which industrial development and national power can be made to grow. The Institute has thus set forth this important mission as its basic objective.[8]

The new institute was organized in sections (physics, chemistry, etc.), each headed by a prominent researcher who was granted considerable autonomy in budgeting and research design. Most of the researchers held university appointments and used Riken funds to pay for younger assistants in their university laboratories; others also had laboratories at Riken itself. Physical planning for the Riken laboratories was delegated to four prominent scientists from the University of Tokyo. The physical section was planned by Hantaro Nagaoka and by Masatoshi Okochi, a professor of ordnance and mechanical engineering, who became head of Riken's industrial group and who founded most of the entrepreneurial firms that came out of it.[9] The chemistry section was entrusted to Kikunae Ikeda, the chemist who had identified and synthesized monosodium glutamate, and Jinkichi Inoue, a professor of applied chemistry.[10]

Riken attracted many of Japan's foremost scientists. Among them were Shin'ichiro Tomonaga, a future Nobel laureate who had studied under Nishina; Umetaro Suzuki, an organic chemist and a co-discoverer of vitamin B1; and Kotaro Honda, a physical metallurgist who invented a path-breaking permanent magnet with four times the coercive force of any existing permanent magnet. Honda was a protégé of Nagaoka at Tohoku Imperial University. The plan that these scientists put together for Riken was a departure from the bureaucratic scientific traditions of Japan's universities. It was a virtual model of the type of institution and organization of research later advocated by Kelly and recommended by the U.S. National Academy of Science's Advisory Group. The funding came from the Imperial family, from leading businesses, and from government subsidies.

Unfortunately, no sooner had Riken been established than it ran into difficulties. When World War I ended, Japan's markets were again flooded with products from the West. A number of financial setbacks for Riken ensued, including unanticipated construction costs, lower-than-anticipated private contributions, and an unexpected rise in the costs of materials and labor. Unable to finance its operations with interest earned on capital investments, as had been planned, Riken ventured into manufacturing.

Between 1923 and 1925, under Okochi's direction, Riken incorporated two companies to manufacture and market synthetically fermented rice wine and vitamin A.[11] This entrepreneurial spirit, combined with a capacity for technological innovation, quickly made Riken Japan's preeminent facility for theoretical and applied scientific research.[12] Despite financial obstacles, the institute established a sound financial base as the number of satellite companies rose from one in 1927 to 60 by 1940. The staff grew from 350 to 750 by 1935, and to 1,800 by 1940.[13] Riken also carried out research on the rates of chemical reactions, the structure of pure inorganic acids, the effect of low temperatures on fluorine, the synthesis of fiber from polyvinyl alcohol, polarity in inorganic compounds, and polymers.[14]

Riken's close alliance with industry helped it both to promote basic research and to substitute domestic for foreign technology. After the Manchurian Incident of 1931, however, its economic base increasingly depended upon military needs. As military contracts increased, so did pressure from the military to abandon basic research. By 1938 more than 80 percent of the Riken's output was war-related.[15]

As the war turned against Japan, the fortunes of Riken declined. By late 1944, Nishina's large cyclotron was inoperable for lack of electricity. At the end of the war, many of the facilities were in shambles from firebombings and the institute was in severe debt. During the occupation, Riken's director, Dr. Okochi, was arrested as a suspected war criminal and was imprisoned for four months. In November 1945 (as mentioned above) army engineers destroyed the cyclotrons that would have provided the basis for renewed research. Riken's "golden years" were over, but the most severe challenge to its survival was still to come. That challenge would be posed by the policy of dissolving *zaibatsu* firms—those large conglomerates of industrial, commercial, and banking interests that seemed opposed to the occupation's goal of economic democratization.

The decision to force economic deconcentration was controversial. It had its critics both within the State Department (which had developed the policy in the Basic Directive for the Occupation of Japan) and within GHQ-SCAP (which had responsibility for implementing the policy through the Antitrust and Cartels Division of the Economic and Scientific Section).[16]

The deconcentration policy was apparently based on a study prepared at the outset of the occupation for the State Department's International Business Practices Branch. The study was carried out by a group of experts that included Corwin Edwards of Northwestern University, a former economic adviser on antitrust matters to the U.S. Attorney General during the New Deal.[17] This report, which described and analyzed the structure of the zaibatsu firms in great detail, concluded that the concentration of Japanese business was significant and that a policy of deconcentration was clearly indicated.[18]

At first this policy was aimed only at the largest major holding companies (Mitsui, Mitsubishi, Sumitomo, and Yasuda), which were required to publicly sell their securities, but it was soon extended through broad anti-monopoly provisions. These were incorporated into a law passed by the Diet, under strong pressure from SCAP, which prohibited trusts, cartels, interlocking corporate control, agreements in restraint of trade, and any other arrangements tending toward monopolies and which established a "Fair Trade Commission" to police business against practices that threatened free and open competition.[19]

By December 1945, SCAP had prohibited the Riken Industrial Company and 325 other firms from declaring dividends or disposing of any funds or assets without its approval. In May 1946, the Antitrust and Cartels Division moved to dissolve the Riken Industrial Company, whose basic purpose, as Nishina argued, was to provide financing for essential research.[20] The loss of funds from the holding company would spell disaster for the research institute.

Only two factors prevented Riken from selling its securities to liquidate its loans and ceasing operations completely. One was the extensive war damage to its facilities. The other was Harry Kelly's desire to see this institute preserved as a model scientific enterprise dedicated to peaceful research.[21]

In May 1946, Kelly informed Brigadier O'Brien that the Cartels Division would try to allow the research institute to survive, although the "method of survival and the role of the Institute in Japanese science" had yet to be determined.[22] Survival depended on two conditions. First, Riken would have to satisfy its creditors, obtain loans to resume research, and pursue profitable research and development. Second, the organizational form would have to satisfy both SCAP and Japanese law.[23]

Nishina's reaction to this was reprinted in a 1978 special issue of the Japanese journal *Shizen* on the history of Riken:

At first I couldn't quite understand what it was all about, but then Kelly had it explained to me by a man from ESS. Riken was to be liquidated because we violated the Anti-Monopoly Law on the grounds that Riken held the stocks of a large number of companies and survived on their dividends, which met the definition of a holding company, thus requiring liquidation.

I repeated my consultations with Kelly, who did everything possible to get it across [to the Cartels Division] that this Institute was indispensable for the rehabilitation of Japan. But at that time, the anti-trust faction in GHQ was obdurate beyond words. Kelly held out for us to the end, and finally talked them into agreeing that Riken was needed for the rehabilitation of Japan. The question then became how Riken should be organized in order to remain in existence.

The corporate status of "zaidan hojin" [a foundation status that was granted to assist specified companies] was eyed with disfavor in GHQ, which was determined to snuff out anything known by that name.[24] In the end, Dr. Okochi and I were called up [to SCAP Headquarters]. Kelly was there too along with two or three others from the anti-trust school. Representatives of banks and finance corporations who said we owed them a lot of money were also present. They proposed reorganizing Riken as a for-profit company assigned to continue Riken's activities, and abandoning the zaidan hojin status. Dr. Okochi appealed to maintain the status, but his pleas were not heard.[25]

Although Kelly had suggested an alternative to a pure for-profit organization (he proposed at one point that Riken be attached to a university), he was not entirely sympathetic to preserving the original zaidan hojin status. Kelly and his colleagues in the Scientific and Technical Division understood the status as an instrument of modernization, but they were critical of practices that allowed Riken to avoid taxes and suspicious of structures established with government collaboration.[26]

The solution to Riken's survival agreed to by Okochi and the creditors at the July meeting was the creation of a new joint-stock company that would be allowed a trial period of six months or more to determine its financial viability. If it failed, its assets would be liquidated to satisfy the remaining creditors.[27] After the decision, the scientists of Riken selected Nishina to lead the reorganization.

In addition to his stature as one of Japan's leading scientists, there were two strategic reasons for Nishina's selection: GHQ trusted him, and he was a capable speaker of English. Okochi had been advised by Prime Minister Yoshida to separate himself from Riken. He resigned in October, and the next day Nishina was nominated by the new board to take over as president.

In the ensuing months, Riken faced a rocky financial situation and numerous bureaucratic hurdles in the wake of GHQ's shifting priorities and administrative muddles in implementing the deconcentration policy. In early 1947, a frustrated Nishina sought Harry Kelly's assistance in securing approval from the Cartels Division for a major loan from the Reconstruction Finance Bank to rebuild Riken's facilities. The approval application had languished at GHQ since the previous November, and rumors began to circulate that it had been destroyed. When Kelly found out that the application was still active he immediately introduced Nishina to Cartels personnel, admonishing that the ability to pursue peaceful research was crucial to the economic rebuilding of Japan. By that time, however, support within GHQ for the survival of Riken had diminished. Complaints from both the Cartels Division and the Finance Division suggested that concerns lingered at GHQ about the excessive concentration of economic power, and intelligence specialists in the Government Section wanted to break up Nishina's laboratory because of its high concentration of nuclear physicists.[28]

Kelly spent many hours arguing for the survival of Riken and for keeping its scientists off of the Government Section's "purge lists," emphasizing that the "old guard" had already left the organization. He even went so far as to claim the intelligence issue in support of Riken's existence. He noted that intelligence was part of his mission and argued that an intact institute with the nuclear scientists working together would be easier to keep an eye on. Yet, although he advanced the security argument for the sake of Riken, Kelly did not believe it. He suggested in an interview many years later that he was chagrined that he had to resort to it.[29] Kelly's approach to surveillance was to expect self-reporting by Japanese scientists and engineers, and he often reiterated his belief that cooperation, not surveillance and policing, was the key to real security.

On April 23, 1947, Kelly finally obtained official SCAP approval for the reestablishment of Riken, and on April 30 the loan was approved. Nishina and Kelly then met frequently on ways to keep Riken solvent in the face of Japan's weak economy. Once Kelly informed Nishina that GHQ would not object to Riken's manufacturing and selling products to keep the institute viable, they discussed what to produce. Nishina decided on penicillin for a number of reasons, including the high demand, the availability of expertise, and the symbolic significance of penicillin as a peaceful application of science. In a newspaper editorial, he wrote:

Modern war, in contrast with the wars of the past that did not make use of science, is ever more ruthless to human life and property, and leads culture—the crystallization of human effort—ever closer to annihilation. . . . Of course, it is also true that science promotes human welfare. Thanks to scientists' wartime efforts, the miracle drug penicillin was born. Those scientists' efforts will probably be remembered forever. In short, science turned to good use will bring about prosperity for humankind; when abused, it will bring devastation.[30]

The attempt to keep Riken alive through the production of penicillin succeeded, although the institution was radically changed and never attained its prewar level of activity. By the summer of 1948, the penicillin operation was paying for itself; later it provided significant support for the entire institute. In ensuing years Nishina and Kelly would each credit the other with saving the institute.

Kelly's Departure

Kelly and Nishina worked well together against formidable odds, both when dealing with GHQ-SCAP and when confronting the tradition-bound institutions of Japanese science. By the end of 1949, however, Kelly knew that his mission was coming to an end. The Science Council of Japan was in place. Riken had been organized under a new name and was functioning at the crossroads of basic and applied science. Through the efforts of the Scientific and Technical Division, Japanese scientists were increasingly in touch with their colleagues, owing to the improved flow of literature, the freedom to present their work at conferences, the improved opportunity to travel, and the contacts initiated by the NAS visits that Kelly had organized. Kelly's three-month mission had extended beyond three years; now it was time to leave Japan to the Japanese. In January 1950 Kelly and his family returned to the United States.

A Bond of Respect and Friendship

Nishina died suddenly and unexpectedly in January 1951. In September 1952, the official peace treaty between the United States and Japan was signed in San Francisco, and the occupation officially ended six months later. Broadcast throughout both countries, the San Francisco ceremony undoubtedly evoked strong feelings in all its listeners, especially those who had worked together to restore Japan. Among them was Sumi Yokoyama, Nishina's former personal assistant, who had joined Nishina at Riken in 1941 and had remained his assistant and confidante through the war and the occupation. As Nishina's widow was ill and unable to care for the family, Yokoyama became a foster mother to Nishina's two sons, Yuichiro and Kojiro.

During the years the Kellys were in Japan, the two families forged a friendship that would endure long after the Americans had left Japan. Both of Nishina's sons eventually pursued graduate study in the United States (Yuichiro in 1953 and Kojiro in 1965) and spent many holidays with the Kellys and their two sons. Harry Kelly often acted as a stepfather to the Nishina sons. He traveled to Japan for Yuichiro's wedding, and when Kojiro married in the chapel at North Carolina State University (where Kelly was Dean of the Faculty) Kelly stood in for the bride's father and escorted her down the aisle.

The bond forged between the families is captured in an exchange of letters between Yokoyama and Kelly on the signing of the Peace Treaty in 1951. On September 5, Yokoyama wrote:

Dear Harry:

This afternoon I listened to the radio to hear Mr. Truman's speech at the opening of the San Francisco Conference. I was very much impressed by his talk in which I could see the true friendship of your country. I could not keep back tears of gratitude thinking what you, Americans, have done for us these six years.

This is what I feel now and I will never forget the kindness your country extended to us. I am sure all Japanese feel as I do now. I do not think that I myself can be of any more use for the better of Japan, but I will do my best to plant and cultivate the seed of peace-loving spirit in the minds of young people around me and help them in realizing the responsibility they have on their shoulders for the future of Japan, which connects directly to the peace of the world.

I am writing you this, because I want (right now) to say . . . "thank you."

Please [convey] my love to Irene, Henry and Tommy.

On September 29, Harry responded:

Dear Sumi,

Irene and I read your letter of thanks with very deep feelings.

We watched the Peace Treaty negotiations on television until two o'clock in the morning. We also felt deep thanks, and we chiefly thought of people like you and Dr. Nishina. Irene and I both want to express our thanks to you for we really feel closer to you and Nishina than anyone else in Japan. We are thankful there are people like you on the other side of the world with whom it is possible to have a very personal friendship. We both want to thank you for letting us be so close to you during this period.

Another thanks I have is for the privilege of being so close to people who were at a very critical stage of their history who showed intelligence, energy and sincerity. I would like to add that if I have become a better man it is because I have worked with Japanese of great stature and for this I give thanks.[31]

Bridging Two Worlds

6

Toward the end of 1949, soon after Kelly informed his friends that he and his family would be returning to the United States, Kunio Senshu, chief of the executive office of the Scientific and Technical Administration Commission, wrote to express the feelings of many of the scientists with whom Kelly had worked over the preceding four years:

Dear Dr. Kelly,

We all deeply regret your return to the United States just at this time of awakening of this country from her postwar stupefaction and the gradual signs of her rehabilitation [when] science and technology are becoming a serious matter of concern among the people at large.

Among your memorable achievements at ESS-SCAP, especially the JSC and STAC owe their birth (wholly) to your pertinent guidance, and we cannot but now admire your foresight.

Comparing Kelly's contributions to Japanese science to those of William S. Clark, an American scientist who had come to teach at the Sapporo Agricultural College in 1876, Senshu said that Kelly's advice would lead to advances in Japanese science and technology similar to those that had followed Clark's departure in the early years of the Meiji Period. With good humor, Senshu reminded Kelly of Clark's departing words to his students: "Boys, be ambitious!"

Harry Kelly returned to the United States with his wife and two young sons in February 1950. Kelly became the head of the regional Scientific Section of the Office of Naval Research (ONR) in Chicago. Set up in 1945 to redirect some of the Navy's research funds from war-related research to basic science, the ONR was one of the first of many institutions established during the first postwar decade to organize American science. The object of the Navy's support was not immediate benefit but, rather, long-term gains that would accrue to the Navy from the general advancement of science. By pumping resources back into university laboratory research that had been neglected during the war, the Navy, according to I. I. Rabi (who had been assistant director of the wartime Office of Scientific Research and Development), "saved the bacon for American science."[1] The ONR's military directors gave free rein to the civilian scientists who were overseeing the research projects. The agency acted as a shadow foundation of the kind Vannevar Bush had proposed and proved to be a training ground for some of the key leaders of the National Science Foundation, which was organized in May 1950. For example, Alan Waterman, a chief scientist at the ONR, became the NSF's first executive director.[2]

From ONR to NSF

Kelly's reputation had preceded his return to the States. By the time the NSF was being established, Alan Waterman had to compete with a State Department science adviser who said that "he was staking a claim on Kelly."[3] In the end, State lost out. In 1951 Kelly accepted the NSF's offer of a job in Washington as Assistant Director for Scientific Personnel and Education. In addition to Kelly, Waterman recruited most of the top NSF staff from among his former colleagues at the ONR.

In *Science—the Endless Frontier,* Vannevar Bush emphasized the renewal of scientific talent as a primary responsibility of his proposed national foundation for scientific research. He believed that "in every section of the entire area where the word 'science' may properly be applied, the limiting factor is a human one. We shall have rapid or slow advance in this direction or in that depending on the number of really first-class men who are engaged in the work in question. . . . So in the last analysis the future of science in this country will be determined by our basic educational policy."[4] Bush and his advisory committee were concerned with two educational goals in the work of the foundation: broadening the base from which students with scientific aptitude and talents could be drawn, and filling the gap the war had created in the supply of young scientists and engineers. The second goal took on even greater urgency with the start of the Korean War in June 1950, which brought large increases in the Defense Department's research budget but which placed heavy strains on the supply of expertise as young scientists were again diverted from research by the draft or by jobs in defense-related industries.

Responding to this scientific shortage, the NSF board and Waterman moved quickly to establish a Fellowship Program to support young students in the natural sciences. The program was set up in the Scientific Personnel and Education Division, which Harry Kelly was about to join. In time, Kelly recruited Bowen C. Dees, the man who had followed Kelly into the Deputy Chief position of the GHQ-SCAP Scientific and Technical Division. In the first year of its operation, the NSF program awarded pre- and post-doctoral fellowships to 575 of the 2,700 applicants. In the first ten years of the program alone, over 12,000 fellowships would be awarded.[5]

Kelly took bold steps to launch educational programs and summer institute programs to improve the skills of college teachers of science. The early models for the institute programs bore the Kelly imprint of independence in implementation. For example, the first conference was sponsored by the University of Colorado, at the NSF's suggestion, but was planned and organized largely by the Committee on Regional Development of Mathematics of the Mathematical Association of America. Allowing professional societies to plan programs avoided the appearance of government control and reflected the NSF's and Kelly's philosophy that science policy should be made by scientists, independent of political and bureaucratic influences.[6] The approach was not unlike the one Kelly had successfully applied in Japan.

The NSF education programs faced a number of challenges that affected the entire country in the 1950s. The landmark Supreme Court decision in *Brown v. Board of Education,* ending school segregation and forbidding the awarding of federal funds to segregated schools, posed a complex problem for the NSF: how to deal with institutions that were in fact segregated and yet were most in need of funds to improve science education. The launching of Sputnik, in 1957, posed another set of challenges, dramatically increasing the interest in and the demand for science education programs throughout the country.[7] Together, these challenges transformed the summer programs for training science teachers from a small number of annual institutes with eminent scientists as lecturers to broad-based programs in all but five states, with many small state schools and liberal arts colleges acting as host institutions.

Kelly's Continuing Ties with Japan

Harry Kelly would remain at the National Science Foundation until 1962. His personal ties with Japan and his commitment to fostering international cooperation between the United States and Japan on scientific affairs remained strong. When Japan assigned a Science Attaché to its embassy in Washington, Naoto Kameyama (who was still president of the Science Council of Japan) wrote to ask Kelly to help introduce the new representative to the world of international science diplomacy. With customary graciousness, Kelly responded to Dr. Kameyama immediately and assured him that he and Bowen Dees would meet the new attaché, Takashi Mukaibo, at the airport. His letter continues: "Irene and I will try to help him get used to American life, and we shall make every effort to see that he is properly introduced around Washington. I was very glad to learn of his past associations with you, and hope that some of your scientific statesmanship will be used to help guide him on his new duties."

Kelly fulfilled his promise, introducing Mukaibo to the scientific establishment in Washington, inviting him to regular meetings and lectures of the National Academy of Sciences, and ensuring that he was well connected with the other science attachés who were being named to Washington. At the outset of his appointment, Mukaibo's primary interests were to promote the Japanese-sponsored Antarctic Observation Project and to gather information on nuclear energy. Many science diplomats had become interested in this topic after President Dwight D. Eisenhower's December 1953 speech to the United Nations, in which the president had urged international cooperation in developing peaceful uses of atomic science. Eisenhower's proposals led to the Atoms for Peace Program, through which the United States shared atomic materials and undertook cooperative research programs focusing on atomic energy with more than 40 countries.[8]

The Mukaibo and Kelly families became friends, and their children often played together. Mukaibo remembered: "One New Year's day, I took Kelly to the Japanese Embassy to partake of mochi—a rice cake eaten traditionally to celebrate the New Year. I remember introducing him to the ambassador, telling him that, unlike many Americans, Kelly was really a Japanese at heart. Political differences aside, tensions between Japan and the United States were kindled by many incidents of misunderstanding and prejudice. Kelly, to his credit, set an example of behavior for others to follow in order to overcome these tensions."

Mukaibo saved Kelly from innocently participating in one of the aforementioned incidents when, in 1954, he advised him to turn down an invitation from the U.S. Navy to set up a liaison office in Japan to investigate the exposure of Japanese fishermen on the ship *Lucky Dragon* to radioactive fallout from U.S. nuclear testing on the Bikini Atoll. Mukaibo gave three reasons for this advice: first, it was inadvisable to go without an invitation from the Japanese themselves; second, given the misgivings of many Japanese scientists about liaisons between science and the military, it was inadvisable for Kelly to participate in a military mission; third, the issue he would be dealing with was extremely volatile in Japan, where people remained very sensitive about radioactive fallout and its consequences.[9] Kelly must have realized that the mission was not compatible with his own objectives of serving science in both countries. He took Mukaibo's advice and stayed in Washington, at least for a while. Soon, however, he was offered another opportunity that was in keeping with his personal beliefs and professional interests.

The U.S.-Japan Committee on Scientific Cooperation

In 1960, a proposal for revisions in the security treaty between the United States and Japan unexpectedly provoked massive demonstrations in Japan.[10] In their wake, both countries saw a need to increase popular support for the alliance. As part of the attempt to bring the countries closer, Edwin O. Reischauer, the U.S. ambassador to Japan, proposed the creation of a U.S.-Japan Committee on Scientific Cooperation.[11] Prime Minister Hayato Ikeda's mission to the United States in June 1961 provided a good opportunity to announce the formation of this committee.[12]

The explicit mission of the Committee on Scientific Cooperation was to develop cooperative research programs that would be carried out by scientists in the two countries. An underlying purpose was to demonstrate that cooperation could yield scientific benefits for the people of both countries.[13] The success of the cooperative research programs would therefore depend not only on the quality of the research the committee supported but also on the ability of the two national delegations to project a spirit of cooperation and goodwill. Kelly, who had built friendships with many prominent Japanese scientists and had also earned the respect of his colleagues in the United States, was the natural choice as the American co-chairman. The Japanese selected Kankuro Kaneshige as their co-chairman.

Kelly and Kaneshige were familiar with each other's working style from the days when the Science Council of Japan was being created. This mutual understanding was crucial to the success of the committee, which was made up of eminent scientists from the two countries, each with his or her own expectations and working style. In this context, the need for people who could smooth over cultural misunderstandings was apparent. For example, when some of the American scientists became impatient at the lengthy and seemingly directionless discussions at some committee meetings, Kelly was able to point out the importance of ceremony in Japanese professional life. He understood that the lengthy discussions were important, if for no other reason, as a symbol of unity.[14]

Some of the other members of the first U.S.-Japan Committee on Scientific Cooperation had, like Kelly and Kaneshige, been involved in rebuilding ties between the American and Japanese scientific communities. Detlev Bronk, who had led the second Science Advisory Group to Japan, represented the National Academy of Sciences. Kelly's longtime friend Seiji Kaya and the then-current president of the Science Council, Kiyoo Wadachi, were also on the committee. These members—known to one another and respected in both countries—provided stability and reassurance.

At first, official representation of the Science Council of Japan on the Committee on Scientific Cooperation was contested by Japanese members who were hostile to the U.S.-Japan Security

Treaty. Some were generally wary of American politics and policies and concerned that further economic, cultural, and scientific collaborations between the two countries would merely serve American political purposes. Since the founding of the Science Council there had been tensions between the liberal and leftist scientists who were often elected to it and the politically conservative governments that held power from 1949 on. The leftist opinions of the scientists were viewed with suspicion by Japan's leadership (and by some Americans—this was the period of "red-baiting" in the United States). In 1954, a scientist working for the Atomic Energy Commission set off something of a furor with an offhand remark to the effect that Takashi Mukaibo, Japan's science attaché, had "leftist ideas." The potential damage to Mukaibo's credibility was defused only when Harry Kelly came to his defense and extracted an apology from the American scientist.

Although red-baiting had been discredited by 1961, relations between the United States and the Soviet Union had deteriorated markedly and anti-leftist sentiment was still very strong. This concerned some members of the Science Council of Japan, but these concerns were at least partially assuaged when the members of the Committee on Scientific Cooperation issued a declaration that it would "limit, both in spirit and in substance, the activities [of the committee] strictly to the advancement of science, knowing that any political good would come about as a by-product."[15] The joint communiqué emphasized that these activities would be "purely scientific and dedicated to the service of humanity and the arts of peace."[16] The Science Council finally agreed to send its president in an ex-officio capacity, expecting him to "fully represent the views of Japanese scientists."

Harry Kelly felt that by the early 1960s the scientific relationship between the United States and Japan required further thought. "Japan had recovered economically," he wrote, "and was able to contribute her share to the advancement of scientific knowledge. It was time to consider a policy of mutual aid, rather than foreign aid, for the advancement of science in both countries."[17]

At the first meeting of the Committee on Scientific Cooperation, held in Tokyo on December 13–15, 1961, the two delegations agreed on three areas of cooperative scientific effort which they believed would benefit both countries: the exchange of scholars, the exchange of more scientific information and materials, and the implementation of specific research programs on the geography and ecology of the Pacific Ocean and on cancer.[18] These joint efforts became the responsibilities of different agencies in the two countries. In Japan, a Coordinating Council presided over by the vice minister of education for science and technology designated individual scientists or groups of scientists to carry out the projects, with funding from the Ministry of Education and administration by the Japan Society for the Promotion of Science. In the United States, the National Science Foundation was made responsible for administering the committee's projects.

Throughout the 1960s Kelly and Kaneshige held the Committee on Scientific Cooperation together and set a congenial tone for its deliberations. Their interpreter for the first session was Masao Yoshida, a son of the former prime minister Yoshida (who, 15 years earlier, had assisted Kelly and Kaneshige during discussions about the Renewal Committee). Kelly and Kaneshige jointly chaired the committee through 1969, when Kelly retired from the post.

Sowing Seeds of Mutual Understanding

Kelly left the NSF in 1962 to become dean of the faculty and then provost at North Carolina State University. During his tenure there he continued to work to bridge the worlds of science and the humanities. A press release announcing his appointment as provost noted: "Because Dr. Kelly is, as provost, the chief educator on the N.C. State campus, science and the humanities will converge on this campus. And because of that, we will have better scientists coming out of N.C. State."[19] John Caldwell, former chancellor of the university, recalled that "what Harry Kelly wanted for the whole world was peace and mutual understanding, and he felt that the academic world should behave the same way and work for the same goals."

At N.C. State, Harry Kelly took pains to involve himself in the education of students, going so far as to act as guarantor for a student loan on at least one occasion. He and Irene often opened their home to Japanese students. One of them, Isao Idota (now president of the Technology Transfer Institute, Inc.), credits Harry Kelly with pointing him in the right direction. Idota came to the United States as a postgraduate student in 1965 not knowing exactly what he wanted to do, but wanting to learn more about the United States. With only a modest introduction via the National Science Foundation, he went to visit the Kellys in North Carolina. The family met him at the airport, welcomed him into their home, and spent long hours discussing U.S.-Japan relations. Encouraged by Kelly, Idota returned to Japan with introductions to Professors Kaya and Kaneshige and with the desire to build a company to facilitate technology transfer and information exchange between the two countries.

Epilogue

Harry Kelly died suddenly, at the age of 67, on February 2, 1976. Late in the afternoon of that day, while Irene prepared the evening meal, he went to their bedroom, saying he needed a nap before dinner. He died in his sleep. His Japanese friends held a memorial service on February 27. In attendance were Professor Asami, who had been with Kelly on his first visit to Hokkaido in 1946; Kelly's friends Tamiya, Kaya, and Kaneshige; Nishina's son Yuichiro; Sumi Yokoyama; Takashi Mukaibo, Japan's first science attaché to the United States; and several of Mukaibo's successors in that post.

The following July, in keeping with the wishes of both the Nishina and Kelly families, Irene brought a portion of Kelly's ashes to Japan for burial in the Tama Cemetery next to the grave of Dr. Nishina. There, a modest tombstone bearing a simple inscription in Dr. Kaya's hand may one day inspire others to wonder who this man was and why he is remembered with such respect in Japan. It says, simply, "Here lies Harry C. Kelly."

In the period since Kelly had left Japan, the nation had entered a new era. The gross national product had increased at an average rate of about 10 percent per year. Research and development, just beginning to reemerge in 1950, was now receiving significant attention and funding from the government. Driven in the early postwar years by the need for recovery and by a motivation to catch up with and surpass the West, Japan's importation of technologies from abroad, coupled with research that had led to innovative applications, had provided a foundation for increasing economic strength.

Since Kelly's death, Japan's economic success has led to another set of challenges. Now that Japan has attained economic superpower status, policy makers and business leaders are launching efforts to stimulate even greater creativity in science and technology. Their emphasis has shifted from research on applications to measures to improve the country's basic research capabilities and to open and share its research facilities with scientists from other countries.[1] In light of this new version of what has come to be called "technodiplomacy," a few scientists and others in Japan who remember the example of Harry Kelly have suggested that it would be useful for the country to reexamine and apply his proven methods of building mutual understanding.[2]

Appendix

Remarks to the Science Council of Japan
on the 25th Anniversary
by Harry C. Kelly

October 25, 1974

Mr. President, officers, and members of the Science Council of Japan, and distinguished guests.

It is a great honor to be invited to this 25th Anniversary of the Council. I should like to add my word of congratulations for your contributions to the advancement of scholarship, the preservation of academic freedom, and the fulfillment of your international scholarly responsibilities.

In his kind invitation, Dr. Ochi, your President, encouraged me to give a brief historical account of my experiences in working with the scholars of Japan during the occupation. I welcome this opportunity to review some of the struggles we, Japanese and Americans, had in searching for an appropriate relationship between science and governments during a difficult period of our history.

At the outset, all of us, Japanese and Americans, were concerned with determining the role of scholars in the strange environment of a military occupation. The presence of civilian science advisers to the military occupation was due to both guilt and fear on the part of the United States. There was a deep feeling of guilt for the use of atomic bombs at Hiroshima and Nagasaki and the terrible human casualties which resulted. Our fear resulted from our knowledge of Japanese scientific capabilities in nuclear research and our ignorance of how far the Japanese had progressed in developing these capabilities. Such were the reasons we first came together—guilt and fear.

The destruction of the Japanese cyclotrons brought to a head the desperate need for a science policy in the occupation. As a result, the late Dr. Gerald Fox of Iowa State University and I were invited to go to Japan as science advisers to the military occupation.

To begin with, we science advisers were without much guidance except for a few directives from Washington attempting to limit and prohibit certain kinds of research. Clearly we had to develop policies and attitudes to guide our relationships with the Japanese. The Japanese scholars, on the other hand, had a similar problem in knowing how to deal with us.

The prohibitions admitted many interpretations and, if interpreted strictly, would have required a significant percentage of all American scientists to ensure inspection and compliance. Further and more importantly, this interpretation had to be done by people, both Japanese and American. Take, for example, the concern for the Death Ray apparatus which had been under experimentation by the Japanese. It was essentially a rather large magnetron whose primary purpose was to direct destructive radar rays at aircraft. The experiment failed—LASERS had not yet been invented. Rumor was that the apparatus was at the University of Hokkaido. But after several days on the campus I could find no trace of it, although I had a hunch that some apparatus was moved in preparation for my visit. When I informed the university officials of my disappointment, they informed me of their dilemma. To be sure, they had such apparatus. The experiment was a failure; but they were concerned about possible incrimination of the university and its faculty and students should it be found on campus. They, therefore, quietly moved the apparatus to a small elementary school some thirty miles out in the country. I was taken to the school and saw the apparatus—but then the dilemma was mine. What was I to do? After discussion with Japanese scholars present, it developed that many of the parts, such as meters, were in great need in laboratories. I decided then to print a big sign to put outside the room—PROPERTY OF HOKKAIDO IMPERIAL UNIVERSITY, INSPECTED BY DR. HARRY C. KELLY, GHQ, SCAP—just in case some investigators accidentally happened on this remote elementary school.

Fortunately, Japanese scholars were quite as eager as we were that wise policies and friendly relationships should be developed. They understood, too, that just as they must be trusted by their government, we, too, had to have the confidence of ours. I have many warm memories of how these personal relationships prevented many possible policy errors. I remember, for example, Dr. Seiji Kaya coming to my home unannounced late one night. I had known Dr. Kaya for about a year, but my conversations with him were always rather formal and always through an interpreter. Dr. Kaya, in perfectly understandable English, apologized for his late visit and explained that he could not find an interpreter on short notice. After a good laugh by both of us at the circumstances, he explained that he had an urgent problem with the national budget for research. An item important to scholarship in Japan was being deleted by budget officials of the occupation who did not understand its significance. Dr. Kaya performed a service to Japanese scholarship and the occupation by preventing a budget error.

As a result of a growing number of such relationships, the Japanese scholars were more willing to share the responsibility of developing policy by self-censorship of research activities during the temporary period of restrictive occupation directives.

This experience emphasizes the absolute need to foster trust and confidence among scholars of all countries. No international organization or inspection systems, for arms control, for example, can be made to work without the free and trusting cooperation of scholars.

It was particularly important that there were a large number of underlying convictions and postulates which were tacitly accepted by all of us. Perhaps most important of these was that the search for truth could be tampered with only at great peril. Others were as follows: Japan, with its great economic difficulties, needed its scholars more than ever in identifying and overcoming obstacles to recovery. Japanese scholars were world leaders in the basic sciences, such as mathematics, genetics, and theoretical physics; but they were inactive in most applied fields. Clearly, with the shortages of raw materials, Japan would have to export its scientific and technical abilities for raw material. International communication and collaboration was important, not only for the advancement of knowledge, but also for aid in developing world conditions of understanding and peace.

These statements, I believe, were immediately accepted by scholars, but not necessarily by governments. Indeed, an important part of the job, both for Japanese and Americans, was to convince their respective governments of the role of scholars in recovery.

Of course, it is impossible to develop a sound and wise policy regarding scholarship without involving the scholars concerned. I shall always be grateful for the gracious and wise help in getting to know and respect Japanese scholars in all fields. I also got some information from our military intelligence agencies. It may be difficult to appreciate now the emotional environment when purges of Japanese scholars for military activities were being considered. Because of the possibility of such purges, my opinion about scientists was requested, and I was allowed to read the dossiers of some leading scientists. One I shall never forget was Dr. Yoshio Nishina. I did not then personally know Dr. Nishina; but, like all students of physics, I had studied and respected his many contributions to atomic physics. Even this unusual source could not hide his greatness as a scholar and humanitarian who during most trying times succeeded in being both a loyal Japanese and a scientist who held the respect of scholars all over the world for his contributions to the advancement of knowledge.

Many scholars contributed to the development of the occupation policies, but I must mention a few like Dr. Nishina and those who were more intimately my teachers and friends like Dr. seiji Kaya, Dr. Kankuro Kaneshige, Dr. Hiroshi Tamiya, Dr. Sakae Wagatsuma, Dr. Riyokichi Sagane, Dr. Naoto Kameyama, and Dr. Juro Horiuchi.

Perhaps the most important advice that I received from these friends when we began discussions of organizations for scholars was that under no circumstance should we attempt to impose an organizational structure on Japanese scholars. They all agreed that the present organizational structure needed a critical review and that it was a most opportune time for change; but all equally urged that whatever was done should be done solely by the Japanese. They agreed that all Japanese scholars should have a voice in this study. Our efforts in the occupation then should be limited primarily in encouraging a self-study and the development of recommendations for change if need be.

The initiation of studies on organizational structures also marked a subtle, but important, change in attitudes between Japanese scholars and the American advisers. The first stage of the occupation was kind of an emergency period when Japanese scholars and American advisers learned to work together in complying with occupation directives and, more importantly, in trying to get affairs back to normal.

The second stage began with discussions on organization, which led to the formation of the Japan Science Council. It was important that this was no longer considered a cooperative venture, but the complete responsibility of the Japanese. The United States had a great interest in the discussions and conclusions, but only as observers. We had, and still have, concerns for the same problems you considered, such as: What should be the relationship between the government and scholars? What should be the balance between academic freedom and the power of the government? Certainly there are limits to academic freedom, but what are they? These are some of the issues considered by the founders of your Council.

It was at this stage that we in the occupation, after discussion with our Japanese friends, decided to invite a group of scholars from the U.S. National Academy of Science to visit Japan. The purposes of the invitation were to get an independent opinion of the occupation policies and especially the attitudes of the occupation toward Japanese scholarship and scholars, and to begin international communication and collaboration with Japanese scholars. Dr. Frank Jewett, President of the National Academy of Science, was most enthusiastic in his response and immediately began arrangements for a distinguished group of scientists, with Dr. Roger Adams as chairman, to visit Japan in August 1947. A second group under Dr. Detlev Bronk visited the following year. Both groups encouraged us in our policies, and international exchanges were started by the U.S. Academy's invitation to the Japan Science Council to visit the United States.

All Japanese scholars who participated in the early stages, from the Preparatory Committee to the Renewal Committee, showed great patience, vigor, wisdom, and willingness to cooperate. I know that every effort was made to ensure that any scholar in any field in any part of Japan, if he were interested, should have his voice heard. It must have been with a great sense of relief and accomplishment that the Renewal Committee, under the chairmanship of Dr. Kankuro Kaneshige, reached its conclusions and recommendations in the General Meeting in March 1948. The law establishing the Science Council of Japan was approved by the Diet in July 1948.

The careful studies made by the Renewal Committee were of special interest to us in the United States, for we had and still have organizational problems ourselves. Indeed, President Truman, during this period, vetoed the first law establishing the National Science Foundation because of his conviction that, if the Foundation were to disburse federal funds, the director should be appointed by and should report to the President of the United States rather than being appointed by and responsive to a group of scholars outside the government.

A few years ago a group of Japanese scholars visited the United States to learn how we were attacking some of our science policy problems. After reporting that we seemed to have the same problems they faced, they told me a story about two frogs, one living in Kyoto and the other in Tokyo, who wanted to see each other's city. They met on a hill half way between with a view of both cities. It was suggested that they stand up to get a better view of their targets. They did, but frogs, having eyes in the back of their heads, saw only the city they came from. After this observation, both decided that there was no point in continuing. Each returned to his home city. But we can't be so backward or short-sighted. Scholars know of the value of observing alternative solutions to problems and the value of mutual attacks on mutual problems.

In many important areas the Science Council of Japan has been unique when compared with similar organizations, at least in my country. I should like to comment on two.

One is that JSC has made a sincere effort to have its membership representative, irrespective of discipline, age, or geography. It may still not be perfect in Japan, but the movement has been in the right direction.

The other area is that of cooperation among the different disciplines of learning. This is a very difficult and increasingly important area of concern. We have come to recognize that many of our challenging scholarly and practical problems are interdisciplinary in nature. These interdisciplinary problems require the cooperation of humanists, social scientists, natural scientists, and applied scientists.

In education as well as in research, an interdisciplinary approach is needed. I remember a graduating senior complaining, after taking many courses in many different fields, that some effort at correlation and unification of these many bits of knowledge should be made. The search for the unity or unities of knowledge in different fields had seemed too difficult for the professors and by default had been left to the student.

Much to the surprise of many Westerners, the framers of JSC took for granted that all disciplines of knowledge should be included in the new council. Article 10 of the Science Council of Japan law, by creating a cultural as well as a natural science department, at least encourages interdisciplinary approaches to common problems.

The need for guarantees of freedom of learning and thought was of great concern to the early founders of the Council. Of special concern was the preservation of academic freedom in its relationships with the government.

This is a grave problem for scholars everywhere. One of our distinguished presidents of the United States, Thomas Jefferson, once described the role of government as treading the fine line between liberty and order.

The nature of the interface between scholars and their patrons is of concern to scholars everywhere, and there are many different approaches to the problem. The founders of JSC, I believe, have made a significant contribution through the invention of the Council.

But organizations are not enough. Needed, too, are wise men of good will on both sides of the interface—men who can appreciate the meaning of and society's need for freedom of thought and the advancement of knowledge, and at the same time appreciate the need for at least a kind of fiscal orderliness and appropriate responsibility to taxpayers and other patrons.

Fortunately, scholars of this calibre came forward during the trying period of the occupation. Sitting as an observer whose primary job was, as stated in the first annual report of the Science Council, to help insulate the deliberations "from all the unreasonable impediments from outside," I ended up with deep respect for the participants. Careful consideration was given to responsibilities of the scholars to their fellow men, as well as to their share in the advancement of knowledge and their responsibilities for ensuring freedom of thought and learning. In the men involved there was nobility as well as wisdom; and they deserve the gratitude of all of us.

Today we take a brief interlude to reflect on problems of the past and to salute those noble men who led us through treacherous times.

But the world's needs for noble scholars like these is even more pressing today. Indeed our survival may depend on our ability and good luck in having men of wisdom, scholarly knowledge, and humanitarianism in positions of leadership and leverage. The importance of problems associated with disarmament, energy, pollution, food, and economic and political stability is being forced upon us. Becoming clear are the responsibilities of scholars in helping attack these problems. Increasingly we begin to realize that they can be wisely attacked only by international cooperation. Indeed, the problems may be so great and grave that they themselves will bring us together on this planet of ours.

Japan also has an ironic responsibility to keep the memory of Hiroshima and Nagasaki fresh, clear, and understood by the whole world. Since the first bomb at Hiroshima, the destructive power of nuclear bombs has increased a thousandfold. Yet we are no closer to effective international agreements on control of this terrible power. Perhaps fear can help bolster wisdom and compassion.

Because the international problems facing us are so great and complex that even with these limitations of our scholars we must sincerely and earnestly search for cooperative solutions. My hope for the future of all of us is that scholar-statesmen of wisdom and humanity, like the ones who came forward during the formation of the Council, will now come forward to help us attack the great problems facing us, and will develop the necessary trust across the interface between scholarship and governments.

Finally, I want to express my own personal gratitude to the scholars of Japan who, in a most unfavorable environment in the beginning, were such great teachers and proved to be such good friends. Trusting friendships can be found by working together on common and significant problems.

Notes

Chapter 1

1. See Edwin O. Reischauer, *Japan: The Story of a Nation,* revised edition (Tuttle, 1976), chapter 11.

2. Theodore Cohen, *Remaking Japan: The American Occupation as New Deal* (Free Press, 1987), p. 100 and n. 2, p. 485.

3. For a thorough bibliography of works in English on the occupation see *The Allied Occupation of Japan, 1945–1952: An Annotated Bibliography of Western-Language Materials,* ed. R. E. Ward and F. J. Shulman (American Library Association, 1974). See also John W. Dower, *Japanese History and Culture from Ancient to Modern Times: Seven Basic Bibliographies* (Wiener, 1986).

4. A notable exception is Cohen, *Remaking Japan.* See also *Americans as Proconsuls: United States Military Government in Germany and Japan, 1944–52,* ed. R. Wolfe (Southern Illinois University Press, 1984).

5. For two important histories of these developments, see Alice Kimball Smith, *A Peril and a Hope: The Scientists' Movement in America, 1945–47* (University of Chicago Press, 1965), and Vannevar Bush, *Science—the Endless Frontier: A Report to the President on a Program for Postwar Scientific Research* (National Science Foundation, 1945). See also Harvey Sapolsky, *Science and the Navy: The History of the Office of Naval Research* (Princeton University Press, 1992).

6. For a history of the early postwar years in Japan, see Shigeru Nakayama, *Science, Technology and Society in Postwar Japan* (Kegan Paul International, 1991).

7. See GHQ-SCAP, "Phase-Out Report," Scientific and Technical Division, Economic and Scientific Section (March 29, 1951), record group 33, National

Archives and Records Administration (NARA), Suitland, Maryland, pp. 1–4. Also available on microfiche: *History of the Non-Military Activities of the Allied Occupation of Japan,* NARA monograph 54.

8. GHQ-SCAP, JSC Serial 18, 3 November 1945, cited in "Phase-Out Report."

9. GHQ-SCAP, Serial 24, 14 December 1945, cited in "History of the Scientific and Technical Division," r.g. 33, Scientific and Technical Division, NARA, box 1, folder 1.

10. Joint Chiefs of Staff (JCS) 1380/15 (JCS Serial 18 of 3 November 1945), in GHQ-SCAP documents, record group 33.

11. Samuel K. Coleman, "Riken from 1945 to 1948: The Reorganization of Japan's Physical and Chemical Research Institute under the American Occupation," *Technology and Culture* 31, no. 2 (April 1990), p. 232.

12. For a perceptive analysis of the role of cyclotrons in the development of nuclear physics as a major research field in several scientific communities in the 1930s, see Charles Weiner, "Cyclotrons and Internationalism: Japan, Denmark and the United States, 1935–1945," in Proceedings of XIVth International Congress of the History of Science 1974 (Tokyo, 1975), on which much of this section is based.

13. Ibid.

14. Coleman, "Riken from 1945 to 1948," p. 230.

15. See John Dower, "Science, Society and the Japanese Atomic-Bomb Project during World War Two," *Bulletin of Concerned Asian Scholars* 10, no. 2 (April–June 1978), pp. 41–54; Charles Weiner, "Retroactive Saber Rattling?," *Bulletin of the Atomic Scientists* 34 (April 1978), pp. 10–12; and Weiner, "Cyclotrons and Internationalism."

16. See John Dower's essay "'Ni' and 'F': Japan's Secret Wartime A-Bomb Research," in his book *Japan in War and Peace* (Free Press, 1993). In this essay, Dower updates and strengthens his findings about the inability of the Japanese to advance atomic weapons research during the war with information that was not available to him in 1978.

17. Yoshio Nishina, "A Japanese Scientist Describes the Destruction of His Cyclotrons," *Bulletin of the Atomic Scientists* 3 (June 1947), pp. 145, 167.

18. Ibid. See also Dower, "Science, Society and the Japanese Atomic-Bomb Project" and Weiner, "Retroactive Saber Rattling?"

19. GHQ-SCAP documents, record group 33, Records of the Scientific and Technical Division, box 2, folder 13, "Cyclotrons," NARA.

20. For an account of the reaction of American scientists to the destruction of the Japanese cyclotrons, see Smith, *A Peril and A Hope,* esp. pp. 352–356.

21. Cited in ibid., p. 354.

22. Ibid.

23. See Daniel S. Greenberg, *The Politics of Pure Science* (New American Library, 1967).

24. Patterson explained that MacArthur had received a confusing telegram from General Leslie Groves (head of the Manhattan Project) ordering the seizure of the cyclotrons. The content of the wire is quoted and cited in Charles Weiner's article "Retroactive Saber Rattling?"

25. Association of Oak Ridge Scientists (AORS) wire to Patterson, December 5, 1945; and Patterson to AORS, December 15, 1945, quoted in Smith, *A Peril and a Hope,* n. 20, p. 355.

26. DuBridge to Jewett, December 31, 1945, file "Org., 1946 (Projects proposed on sponsorship of restoration of Japanese cyclotrons)," in Central Policy File of archives of National Academy of Sciences, Washington (hereafter "NAS Archives").

27. Royall to Jewett, n.d., received January 28, 1946, file "Org., 1946," NAS Archives.

28. Jewett to Lee DuBridge, January 28, 1946, file "Org., 1946," NAS Archives.

29. Harry C. Kelly, interview by Charles Weiner, 77–117 (1975), Institute Archives and Special Collections, MIT Libraries, Cambridge, Massachusetts (hereafter cited as "Kelly, Weiner interview"). Unless otherwise noted, all biographical references are taken from this interview.

30. See Smith, *A Peril and a Hope,* esp. chapter 2.

31. Harry C. Kelly, commencement address, Pembroke State University, June 1970, in archives of Kelly's papers at North Carolina State University, Raleigh (hereafter "Kelly Archives").

32. Henry Kelly (son of Harry C. Kelly), interview by Joanne Kauffman, August 1991, Washington.

33. Kelly, Weiner interview.

34. For a history of the Rad Lab see Henry E. Guerlac, "Radar in World War II," in *The History of Modern Physics 1800–1950,* volume 8 (American Institute of Physics and Tomash Publishers, 1987). See also *Five Years at the Rad Lab* (Massachusetts Institute of Technology, 1947; available in archives of MIT).

35. Kelly, Weiner interview.

36. Ibid.

37. Ibid.

Chapter 2

1. Harry C. Kelly, quoted in Philip M. Boffey, "Harry C. Kelly: An Extraordinary Ambassador to Japanese Science," *Science* 169, July 31, 1970, p. 452.

2. Kelly, Weiner interview.

3. Ibid.

4. See Cohen, *Remaking Japan*.

5. John O'Brien, memorandum to "Executive Officer E&S Section," GHQ-SCAP, November 14, 1945, in r.g. 33, records of Economic and Scientific Section, NARA.

6. Kelly, Weiner interview.

7. Handbook of the Scientific and Technical Division (GHQ-SCAP document, Tokyo, 1946), chapter II, in box 72.24 of Kelly Archives.

8. Drs. Fox and Kelly to Chief, ESS, "Reports, Overlapping Functions, Suggested Consolidation, Questions," memorandum of 14 February 1946, and W. F. Marquat, 22 February 1946, handwritten response, in GHQ-SCAP documents, records of Scientific and Technical Division, NARA.

9. Harry Kelly to Irene Kelly, March 1946, Kelly Archives.

10. Harry Kelly to Irene Kelly, January 14, 1946, Kelly Archives.

11. While all the universities suffered from the economic hardships facing the country, few university laboratories were extensively damaged. See GHQ-SCAP document "Natural Science Research in Leading Japanese Universities," June 1949, box 72.24, Kelly Archives, Of the major universities, only Nagoya suffered extensive (75%) damage from bombing; damage to other major universities ranged from 0 to 12%.

12. Kelly, Weiner interview, p. 24.

13. GHQ-SCAP documents, SCAPIN 984, 25 May 1946, "Memorandum for the Imperial Japanese Government," Records of the Scientific and Technical Division, ESS, NARA.

14. Harry C. Kelly, "Science in Japan," in The United States and the Far East, proceedings of a conference co-sponsored by Duke University and

the American Assembly of Columbia University, Durham, June 5–8, 1957, p. 30 (available in Kelly archives).

15. Harry C. Kelly, "United States-Japan Scientific Cooperation," in *Science in Japan* (American Association for the Advancement of Science, 1965), p. 463.

16. Minojima Takashi, 1968, "Seinenki ni Tasshita Oyo Denki Kenkyu-jo ni Yoseru" ("Reflections on a Research Institute of Applied Electricity That Has Come of Age"), cited in an unpublished paper by Sam Coleman of the University of Oregon ("Harry Kelly and the Early Years of Japanese Science Reconstruction under the Occupation," mimeograph prepared for Washington and Southeast Regional Seminar on Japan, April 20, 1985).

17. Kelly, "United States-Japan Scientific Cooperation," p. 463.

18. Harry C. Kelly, "A Survey of Japanese Science," *Scientific Monthly* 68, January 1949, pp. 42–51; National Research Council of Japan, n.d., "Notes on Activities of Learned Societies and Research Institutes for Natural Sciences," box 73.99, Kelly Archives.

19. GHQ-SCAP documents, Phase-out Report of the Scientific and Technical Division, Economic and Scientific Section, 29 March 1951, Records of the Scientific and Technical Division, ESS, NARA.

20. For an American analysis of the wartime and immediate postwar conditions of the Japanese learned societies, see GHQ-SCAP documents Phase-Out Report (n. 19 above), p. 5; and Science and Technology in Japan (Report No. 10, Economic and Scientific Section/Scientific and Technical Division, Tokyo, October 1947). The latter is available at the Harvard-Yenching Institute in Cambridge, Massachusetts.

21. Kelly, "A Survey of Japanese Science," p. 47.

22. For discussion of the democratic scientists' movement, see Shigeru Nakayama, *Science, Technology and Society in Postwar Japan* (Kegan Paul International, 1991), esp. pp. 14–46.

23. Harry Kelly to Irene Kelly, n.d., Kelly Archives.

24. Ibid.

25. Boffey, "Harry C. Kelly" (n. 1 above), pp. 451–452.

26. Kelly, "United States-Japan Scientific Cooperation," p. 463.

27. Ibid, p. 464.

28. Ibid.

29. In recounting the story, Coleman (see n. 16 above) cites interviews with Seiji Kaya and Shigenori Kinoshita and the article "Takemi Taro-sensei no

Omoide; Tagai ni Jinmyaku o Dashiatte Kyoryoku: Kaya Seiji Ooi ni Kataru" ("Memories of Dr. Taro Takemi; Cooperating by Pooling our Networks: Seiji Kaya Recalls Much") (*Clinic,* April 15, 1984, pp. 52–55).

30. Hiroshi Tamiya, "Nihon Gakujutsu Kaigi Soritsu 10-shunen o Mukaete" ("Saluting the Tenth Anniversary of the Science Council of Japan"), *Gakujutsu Geppo* 12, no. 2 (1959), pp. 5–6, cited in Coleman, "Harry Kelly and the Early Years of Japanese Science Reconstruction under the Occupation" and in Shigeru Nakayama, "The American Occupation and the Science Council of Japan," in *Transformation and Tradition in the Sciences: Essays in Honor of I. Bernard Cohen,* ed. E. Mendelsohn (Cambridge University Press, 1984).

31. See Kiyonobu Itakura and Eri Yagi, "The Japanese Research System and the Establishment of the Institute of Physical and Chemical Research," in *Science and Soceiety in Modern Japan,* ed. S. Nakayama, D. Swain, and E. Yagi (MIT Press, 1974).

32. Tamiya, "Saluting the Tenth Anniversary of the Science Council of Japan," pp. 5–6, cited in Coleman (note 30 above).

33. Nakayama, "The American Occupation and the Science Council of Japan," p. 358 and n. 5.

34. Hantaro Nagaoka (1865–1950) was a pioneer of theoretical physics in Japan. In 1903 he presented a model of atomic structure that arranged electrons ouside a central, positively charged nucleus to the Tokyo Mathematical-Physical Society. While the model aroused interest in Europe, reaction to it in Japan was more muted (James Bartholomew, *Formation of Science in Japan* (Yale University Press, 1989), p. 193). One reason for this cool reception—a feudalistic atmosphere that prevailed in academic circles and suppressed new activities—is illustrative of the problems that the Japanese science reformers (including Nagaoka's son) who were willing to work with Kelly wished to address. According to Mituo Taketani ("Methodological Approaches in the Development of the Meson Theory of Yukawa in Japan," in *Science and Society in Modern Japan,* ed. Nakayama et al.), "When Nagaoka presented his atomic model before a meeting of the Physical Society in Japan, prestigious professors charged that such a theory was metaphysics, not science. Nagaoka was so discouraged that he turned to the study of magnetism."

35. From Charles Weiner, "Cyclotrons and Internationalism: Japan, Denmark, and the United States, 1935–1945," in Proceedings of the XIVth International Congress of the History of Science, 1974 (Tokyo, 1975).

36. "Dr. Kelly's Legacy," February 11, 1986, Asahi Television Network, Tokyo.

37. GHQ-SCAP documents, "Memorandum of Conference, 19 April, 1946," Lt. Col. H. von Kolnitz, Dr. Kelly, Major Slagle, Lt. Commander Crofts, Records of the Scientific and Technical Division, ESS, NARA.

38. Tamiya, "Saluting the Tenth Anniversary of the Founding of the Science Council of Japan."

39. Ibid.

40. GHQ-SCAP documents, "Preparatory Meeting for the Japan Association of Science Liaison," June 6, 1946, in Records of the Scientific and Technical Division, ESS, NARA, RG 33, box 7399, file 4.

41. Kelly, "A Survey of Japanese Science," p. 47.

42. GHQ-SCAP documents, "Meeting of Japan Association of Science Liaison," July 9, 1946, in Records of the Scientific and Technical Division, ESS, NARA, box 7399, file 3.

43. GHQ-SCAP, "Phase-Out Report," p. 6.

44. GHQ-SCAP document, "Reorganization of Science and Technology in Japan," in History of the Non-Military Activities of the Occupation of Japan, n.d. (available on microfiche: no. 54, NARA)

45. GHQ-SCAP documents, "Meeting of Japan Association of Science Liaison," July 9, 1946, Records of the Scientific and Technical Division, ESS, record group 33, NARA.

46. GHQ-SCAP document, "Reorganization of Science and Technology in Japan."

47. GHQ-SCAP documents, "Conversation with Professor Tamiya, June 22, 1946," Records of the Scientific and Technical Division, NARA, box 7399, file 3.

48. Quoted in Nakayama, "The American Occupation and Science Council of Japan," p. 356. Nakayama points out (n. 4, p. 369) that the letter was sent shortly before Kelly's return for the summer to the United States, where he hoped to contact research institutions about the Japanese problems. According to Nakayama, Kelly had argued earlier that the case would be presented best by Japanese scientists themselves and had tried to persuade Tamiya to accompany him. But Tamiya was unwilling, arguing that the plan was premature and not representative of all Japanese scientists.

49. GHQ-SCAP document, "Conversation with Professor Tamiya," June 22, 1946, in Records of the Scientific and Technical Division, r.g. 33, NARA.

50. Ibid.

51. Ibid.

52. Kaya, S., Tamiya, H. and Sagane, R., letter to National Research Council, July 11, 1946, in central policy file, NAS Archives, file "Organizations, 1947."

53. Ibid.

54. Article 9 of chapter II of the Constitution of Japan reads:

Aspiring sincerely to an international peace based on justice and order, the Japanese people forever renounce war as a sovereign right of the nation and the threat or use of force as means of settling international disputes.

In order to accomplish the aim of the preceding paragraph, land, sea, and air forces, as well as other war potential, will never be maintained. The right of belligerency of the state will not be recognized.

55. Kelly, "A Survey of Japanese Science," pp. 45–46.

56. GHQ-SCAP, "Reorganization of Science and Technology in Japan." Although a shift in emphasis took place, spot checks of scientific research continued. A December 1947 memo from Kelly to the Chief of the Fundamental Research Branch notes that the Chief of Division (Australian Brigadier General John W. O'Brien) ordered that "visits should be made to each major university every six months if practicable and that in order to achieve this aim, it is preferable, if necessary, for smaller teams or individuals to make these visits rather than one larger group with a longer interval between the visits" (Records of the Scientific and Technical Division, ESS, NARA).

57. See correspondence files in Kelly Archives; see also Kelly, Weiner interview.

58. See files under "Program for a Self-Supporting Japanese Economy," in GHQ-SCAP, Records of the Economic and Scientific Section, r.g. 33, box 7381, file 7, and box 5982, file 2, NARA.

59. Harry Kelly to Irene Kelly, n.d. (1946), Kelly Archives.

60. Bowen C. Dees, interview, "Dr. Kelly's Legacy" (n. 36 above).

Chapter 3

1. On the role of scientists in the political struggle over control of atomic energy, see Smith, *A Peril and a Hope*.

2. See Daniel S. Greenberg, *The Politics of Pure Science* (New American Library, 1967), esp. pp. 96–125.

3. On this period in the history of science-government relations see Don K. Price, *Government and Science: Their Dynamic Relation in American Democracy* (New York University Press, 1954), pp. 32–65, and Greenberg, *Politics of Pure Science*.

4. Bush, *Science — the Endless Frontier*. See also Harvey Sapolsky, *Science and the Navy: The History of the Office of Naval Research* (Princeton University Press, 1992).

5. See D. S. Greenberg, "The National Academy of Sciences: Profile of an Institution," *Science,* April 14, 21, and 28, 1967.

6. GHQ-SCAP, History of the Non-Military Activities of the Allied Occupation of Japan, film S1040 (MP65–4), roll 10, and referenced documents therein, NARA.

7. GHQ-SCAP, Scientific and Technical Division, Economic and Scientific Section, "Phase-Out Report," 1951, p. 4, NARA. This report notes parenthetically that "although a few German-origin developments were introduced into Japan during the war years, the total number of such ideas was so small as to make very little impression on the country's overall war effort or industrial structure."

8. Ibid., p. 16.

9. Kelly, "Science in Japan," p. 32.

10. Proceedings of the Japanese Diet, 28 September 1946, quoted in GHQ-SCAP, "Reorganization of Science and Technology in Japan (1945 to September 1950)," in History of the Non-Military Activities of the Allied Occupation of Japan (n. 6 above), pp. 72–73.

11. Ibid., p. 43.

12. Kelly, "A Survey of Japanese Science," p. 48.

13. GHQ-SCAP, "Reorganization of Science and Technology in Japan."

14. Ibid.

15. Kelly, "A Survey of Japanese Science."

16. Mitorawa Toshiyoshi, History of the Japanese Science Council, draft, Tokyo University, n.d., in files of GHQ-SCAP, file "Japanese Science Council," NARA; Kelly, "A Survey of Japanese Science," p. 48.

17. Ibid.

18. Ibid.

19. For a discussion of the Minka movement and reasons for its eventual decline, see Shigeru Nakayama, *Science, Technology and Society in Postwar Japan* (Kegan Paul International, 1991), esp. chapter 2.

20. Ibid., chapter 2.

21. Ministry of Education, Scientific Education Bureau, Tokyo, March 20, 1947, cited in Kelly, "A Survey of Japanese Science," p. 48.

22. Frank Jewett to Harry Kelly, June 11, 1947, NAS Archives.

23. Harry Kelly to Frank Jewett, May 28, 1947, NAS Archives.

24. Roger Adams to Frank Jewett, n.d., NAS Archives.

25. Quoted in Tuge Hideomi, *Historical Development of Science and Technology in Japan* (Kokusai Bunka Shinkokai, 1961), pp. 152–153.

26. GHQ-SCAP document, Reorganization of Science and Technology in Japan: Report of the Scientific Advisory Group of the National Academy of Sciences, United States of America, Tokyo, August 28, 1947, esp. pp. 14–20 (box 7423, file "Japanese Science Council," NARA; also in file "International Relations—Japanese Science Council," NAS Archives. (See also box 7429, NARA, "Visiting Experts.")

27. Roger Adams to Alfred Richards, September 4, 1947, NAS Archives.

28. Roger Adams, address delivered at inauguration ceremony of Renewal Committee, August 25, 1947, NAS Archives.

29. Kelly, "Survey of Japanese Science," p. 49.

30. The Science Council of Japan Law, Law No. 121, July 10, 1948, in GHQ-SCAP documents, History of the Non-Military Activities of the Occupation of Japan, appendix 7.

31. GHQ-SCAP documents, History of the Non-Military Activities of the Occupation of Japan, p. 89.

32. The Science Council of Japan Law, Law No. 121 of 19 July 1948, translated by Yamashiro G., in GHQ-SCAP files, record group 33, box 7399, file 12, NARA.

33. Ibid.

34. Ibid.

35. Detlev W. Bronk, "Report on U.S. Scientific Mission to Japan, 1948," recorded during Annual Meeting of National Academy of Sciences, April 26, 1949, in file "Organization—1949, NAS Science Advisory Group in Japan," NAS Archives.

36. I. Rabi to H. Kelly, 8 December 1948, "Limited Importation of Japanese Scientific Personnel to the United States," in GHQ-SCAP, Records of the Scientific and Technical Division, ESS, record group 331, box 1, file 11, NARA.

37. I. Rabi to Gen. W. F. Marquat, December 10, 1948, "The Use of Japanese Research Facilities as an Advanced Base in the Event of Acute Emergency in the Far East," in GHQ-SCAP, Records of the Scientific and Technical Division, ESS, record group 331, box 1, file 11, NARA.

38. GHQ-SCAP document, n.d., "Review of the Third Year of Occupation Science and Technology," Records of the Scientific and Technical Division, ESS, box 7393, file 6, NARA.

39. Law for the Establishment of the Agency of Industrial Science and Technology, Law No. 207, August 1, 1948, in GHQ-SCAP, "Reorganization of Science and Technology in Japan." See also Tuge, *Historical Development of Science and Technology in Japan* (n. 25 above), p. 171.

40. See Johnson, *MITI and the Japanese Miracle,* esp. chapter 5.

41. GHQ-SCAP documents, n.d., "1949 Annual ESS Historical Report," p. 10, r.g. 33, box 7420, NARA.

42. GHQ-SCAP document, "Review of the Third Year of Occupation Science and Technology," p. 6. For a history of MITI see Johnson, *MITI and the Japanese Miracle.*

43. GHQ-SCAP documents, "History of the Non-Military Activities of the Occupation of Japan," p. 73

44. For a biography of Yoshida, see John P. Dower, *Empire and Aftermath, Yoshida Shigeru and the Japanese Experience 1878–1954* (Council on East Asian Studies, 1979), esp. chapters 8 and 9.

45. See Johnson, *MITI and the Japanese Miracle,* p. 189.

Chapter 4

1. See Shigeru Nakayama, "The American Occupation and the Science Council of Japan," in *Transformation and Tradition in the Sciences,* ed. E. Mendelsohn (Cambridge University Press, 1984), pp. 353–369; and Tuge, *Historical Development of Science and Technology in Japan,* chapter 6. Both accounts point to the considerable influence of GHQ personnel in the formulation of the plan that led to the creation of the Science Council of Japan.

2. Harry C. Kelly, "Address by Dr. H. C. Kelly at Inauguration of Science Council of Japan," 21 January 1948, GHQ-SCAP documents, Records of the Scientific and Technical Division, ESS, NARA.

3. Nishibori received Japan's Deming Prize in 1954 for his contributions to the diffusion of statistically based quality control practices. He returned to the Kyoto University faculty from 1956 to 1958 and later joined the Japan Atomic Energy Research Institute, where he participated in nuclear research.

4. Nishibori Eizaburo, unpublished memoirs, n.d., quoted in H. Yoshikawa, *Science Has No National Boundaries* (Mita, 1987).

5. See Nakayama, "The American Occupation and the Science Council of Japan," for a discussion of the opposing points of view on reform.

6. Naoto Kameyama, "The Science Council of Japan," Tokyo, March 1, 1950, in GHQ-SCAP documents, Scientific and Technical Division, ESS, record group 33, box 7420, NARA.

7. Nakayama, "The American Occupation and the Science Council of Japan," p. 367 (emphasis added).

8. The Science Council of Japan Law, Law No. 121, July 10, 1948, cited in Nakayama, "The American Occupation."

9. See GHQ-SCAP documents, "Pacific Science Congress," Scientific and Technical Division, ESS, record group 33, box 7417, file 24, NARA.

10. Ibid.

11. GHQ-SCAP documents, "1949 Annual ESS Historical Report, Scientific and Technical Division," n.d., Scientific and Technical Division, ESS, record group 33, NARA.

12. Science Council of Japan, Annual Report 1949–1950, p. 20, Kelly Archives, box 72.25.

13. Johnson, *MITI and the Japanese Miracle,* p. 209.

14. Science Council of Japan, Annual Report 1949–1950, p. 16.

15. Ibid.

16. See Nakayama, "The American Occupation and the Science Council of Japan."

17. GHQ-SCAP, History of the Non-Military Activities of the Allied Occupation of Japan.

18. Science Council of Japan, Annual Report 1949–1950, p. 23.

19. Ibid.

20. GHQ-SCAP documents, "Report of the Renewal Committee," Scientific and Technical Division, ESS, record group 33, box 7399, file 2, NARA.

21. GHQ-SCAP documents, "History of Scientific and Technical Division, July 1946–Mar. 1950," Scientific and Technical Division, ESS, record group 33, box 7412, NARA.

22. See Nakamura Teiri, "Marxism and Biology in Japan," in *Science and Society in Modern Japan,* ed. S. Nakayama et al. (MIT Press, 1974). At this time in Japan the Atomic Bomb Casualty Commission had also launched a genetics project to study the long-term biological effects of the bombing of Nagasaki and Hiroshima. The project was an opportunity for increased cooperation between American and Japanese scientists, many of whom were engaged in the effort based at Hiroshima. For an analysis of the project see John Beatty, "Genetics in the Atomic Age," in *The Expansion of American Biology,* ed. K. R. Benson, J. Maienschein, and R. Rainger (Rutgers University Press, 1991).

23. Despite his differences with members of the Renewal Committee and the subsequent organization of the Science Council, Horiuchi was one of the scientists to accompany the first Antarctic expedition.

24. GHQ-SCAP documents, "1949 Annual ESS Historical Report," p. 8, Scientific and Technical Division, ESS, record group 33, box 7412, NARA.

25. Ibid.

26. Ibid., p. 19.

27. Ibid.

Chapter 5

1. Kelly, Weiner interview.

2. GHQ-SCAP, History of the Non-Military Activities of the Allied Occupation of Japan. See also Records of the Scientific and Technical Division, ESS, boxes 7410, 7416, 7417, file 3, "Radioisotopes."

3. GHQ-SCAP document, Report of Accomplishments 1949, n.d., p. 12, box 7412, NARA. See also box 7410, file 3, "Radioisotopes."

4. Ibid.

5. This section is based largely on the work of Samuel Coleman. See especially his article "Riken from 1945 to 1948: The Reorganization of Japan's Physical and Chemical Research Institute under the American Occupation," Technol-

ogy and Culture 31, no. 2 (April 1990), pp. 228–250. For the GHQ-SCAP records on Riken, see box 7416, file 21, "Riken," NARA.

6. Itakura Kiyonobu and Yagi Eri, "The Japanese Research System and the Establishment of the Physical and Chemical Research Institute," in *Science and Society in Modern Japan,* ed. S. Nakayama et al. (MIT Press, 1974), pp. 182–197.

7. "Rikagaku Kenkyujo setsuritsu ni kansuru soan" ("Draft for the establishment of the Institute of Physical and Chemical Research"), *Toyo gakugei zasshi* 32, July 1915 (cited in Itakura and Yagi, "The Japanese Research System").

8. Quoted in Michael A. Cusumano's article "Scientific Industry: Strategy, Technology, and Entrepreneurship in Prewar Japan," in *Managing Industrial Enterprise: Cases from Japan's Prewar Experience,* ed. W. D. Wray (Harvard University Press, 1989). Focusing on the industrial side of Riken enterprises, Cusumano provides insight into the institute's contributions to technology development and entrepreneurial management, which he maintains helped to reduce Japan's reliance on foreign engineers and technology imports from the West prior to World War II.

Focusing on the scientific research facilities developed during the same period, Itakura and Eri came to a very different conclusion in their 1957 article prepared for the *Journal of the History of Science, Japan* (no. 41). They claimed that "while the Institute did signify a step beyond earlier conditions, its failure to achieve the scale of activities which the original proposal claimed as justification for its establishment bespeaks the absence of any basic change in conditions." They continue, "These facts also constitute a direct indication that Japanese technology was still unable to shed its colonial-style dependence on the West." (This is quoted from Itakura and Yagi, "The Japanese Research System," p. 195.)

9. For an excellent account of Okochi's role in Riken's decision to invest in commercial manufacturing ventures and in promoting an entrepreneurial "scientific industry" in Japan, see Cusumano, "Scientific Industry."

10. Itakura and Yagi, "The Japanese Research System," p. 190.

11. Cusumano, "Scientific Industry", p. 275.

12. Ibid., p. 270.

13. Ibid., p. 276.

14. Coleman, "Riken from 1945 to 1948," p. 231 and n. 11.

15. Ibid.

16. See Eleanor Hadley, "Deconcentration in Japan," in *Americans as Proconsuls,* ed. R. Wolfe (Southern Illinois University Press, 1984), p. 152 and n. 8, p. 479. Hadley notes that the controversy within GHQ-SCAP centered on the opposing positions of two Chiefs of Section, General Charles A. Willoughby, who headed G-2 (Intelligence Section), and General Courtney Whitney, head of the Government Section, G-3. Concerned about a growing communist threat, Willoughby supported preservation of the conservative elements in Japanese society and opposed radical modification of existing Japanese institutions. Whitney supported economic deconcentration as well as other institutional reforms. For a personal account of the dissension within GHQ, see Charles A. Willoughby, *Shirarezaru Nihon Senryo: Uirobii Kaikoroku* [*Unknown History of the Japanese Occupation*] (Banmachi Shobo, 1973). This book was published posthumously and only in Japanese. In a 1984 article, Eleanor Hadley quotes from Willoughby's memoir the following passage, which gives a sense of the controversy and the tensions within SCAP Headquarters over the deconcentration and other reforms aimed at liberalizing Japan's institutions:

SCAP had embraced with an almost wild enthusiasm the trust-busting ideals that already commended themselves so powerfully to the antitrust division of the Department of Justice in Washington. Some 260 Japanese companies, including some of the most tremendous industrial concerns, had been designated as 'excessive concentrations of economic power. . . . The companies, meanwhile, existed in a state of uncertainty which could not help but interfere seriously with initiative and confidence of management. The ideological concepts on which these measures rested bore so close a resemblance to Soviet views about the evils of 'capitalist monopolies' that the measures themselves could only have been eminently agreeable to anyone interested in the future communization of Japan. Their relation to the interests of Japanese recovery was less apparent.

For a counterpoint to Willoughby's views and an analysis of the deconcentration program, see Eleanor Hadley, *Antitrust in Japan* (Princeton University Press, 1970). For further discussion of the deconcentration program, see Kazuo Kawai, *Japan's American Interlude* (University of Chicago Press, 1960), chapter 8.

17. Kazuo, *Japan's American Interlude,* p. 142.

18. Ibid., p. 140.

19. Ibid., p. 143.

20. GHQ-SCAP document, Notes of a meeting between Kelly and a representative of the Cartels Division, April 22, 1946, Records of the Scientific and Technical Division, ESS, box 7416, file "Riken," NARA. See also Coleman, "Riken from 1945 to 1948."

21. Coleman, "Riken from 1945 to 1948," pp. 332–333.

22. GHQ-SCAP document, H. C. Kelly memo, May 11, 1946, Records of the Scientific and Technical Division, ESS, box 7416, file "Riken," NARA; cited in Coleman, "Riken from 1945 to 1948," p. 235, n. 34.

23. Coleman, "Riken from 1945 to 1948," pp. 235–236.

24. *Zaidan hojin* status conferred prestige and considerable financial benefits to Riken, including tax-free income, eligibility for government subsidies, and tax exemptions for contributors. Okochi's plea to allow Riken to retain this status under a new organizational structure was rejected by GHQ, which objected to categorizing Riken as a "philanthropic organization."

25. "Nishina Kenkyu-shitsu no Ogon Jidai: Saikurotoron to Sono Shuhen" ("The golden age of the Nishina Laboratory: Cyclotrons and their environs"), *Shizen* 12 (December 1978), quoted in Coleman, "Riken from 1945 to 1948."

26. GHQ-SCAP document, "Foundational Juridical Persons—Zaidan Johin," 1947, Records of the Scientific and Technical Division, ESS, box 7420, NARA; cited in Coleman, "Riken from 1945 to 1948," pp. 237–238 and n. 40, p. 238.

27. Coleman, "Riken from 1945 to 1948," p. 237.

28. Kelly, Weiner interview, p. 71.

29. Ibid.

30. Quoted in Coleman, "Riken from 1945 to 1948," p. 245 and n. 81.

31. Correspondence in Kelly archives, North Carolina State University.

Chapter 6

1. Quoted in Daniel Greenberg, *Politics of Pure Science* (New American Library, 1967), p. 134.

2. See Harvey Sapolsky, *Science and the Navy: The History of the Office of Naval Research* (Princeton University Press, 1992).

3. Quoted in J. Merton England, *A Patron for Pure Science* (National Science Foundation, 1982), p. 133.

4. Alan T. Waterman, introduction to Bush's *Science—the Endless Frontier*, p. xv.

5. Ibid., p. xvi.

6. England, *Patron of Pure Science*, p. 238.

7. Ibid., p. 245.

8. See Eugene Skolnikoff, *Science, Technology and American Foreign Policy* (MIT Press, 1967), p. 27.

9. Kelly, Weiner interview, p. 80.

10. See George Packard, *Protest in Tokyo: The Security Treaty Crisis of 1960* (Princeton University Press, 1966).

11. Skolnikoff, *Science, Technology and Foreign Affairs,* p. 148.

12. See statement of Prime Minister Ideka Hayato and President John F. Kennedy, June 26, 1961, quoted in The United States-Japan Committee on Scientific Cooperation: The Thirty Year Report 1961–1990 (revised draft), November 2, 1992 (Washington: National Science Foundation, 1992), p. 6.

13. Skolnikoff, *Science, Technology, and American Foreign Policy,* p. 149.

14. Philip M. Boffey, "Harry C. Kelly: An Extraordinary Ambassador to Japanese Science," *Science* 169 (July 31, 1970), p. 452.

15. United States-Japan Committee on Scientific Cooperation, The Thirty Year Report 1961–1990.

16. Department of State, 1961, "Joint Communiqué of the U.S.-Japan Committee on Scientific Cooperation, Dec. 15, 1961."

17. Harry C. Kelly, "U.S.-Japan Scientific Cooperation," in *Science in Japan* (Washington: American Association for the Advancement of Science, 1965), p. 464.

18. Department of State News Letter, "U.S. and Japan Cooperate in Scientific Ventures," n.d., in Kelly Archives.

19. Reported in *Raleigh Times,* August 9, 1969.

Epilogue

1. Japan Science and Technology Agency, "Science and Technology," *White Papers of Japan, 1988/1989* (Japan Institute of International Affairs, 1990).

2. "Dr. Kelly's Legacy," Asahi Network Television, February 11, 1986.

Index

Adams, Roger, 50, 52–53, 56
Agency of Industrial Science and Technology. *See* Science and Technology Agency
Antarctic Observation Project, 99
Anti-Monopoly Law, 88
Atomic Energy Act (U.S.), 80
Atomic Energy Commission (U.S.), 80–83
Atomic energy, research on, 5, 43
Atomic Secrecy Agreement, 8
Atoms for Peace program, 99

Bennett, Merrill, 50–51
Bronk, Detlev, 56, 101
Bush, Vannevar, 4, 44, 97

Committee on Scientific Cooperation, 101–103
Compton, Karl Taylor, 7, 22
Constitution, of Japan, 40–41
Coolidge, W. D., 50–51
Cyclotrons, destruction of, 5–12, 15, 87

"Death ray" research, 2, 24
Dees, Bowen C., 97
DuBridge, Lee, 10–11

Economic and Scientific Section, 2, 45–46
　Antitrust and Cartels Division of, 88, 90
　Atomic Energy Control Division of, 9
　Industrial Division of, 8
　Natural Sciences Division of, 62
　Scientific and Technical Division of, 8, 20–22, 42, 46, 62, 70, 75, 81, 92
　Special Projects Unit of, 5
Edwards, Corwin, 87
Eisenhower, Dwight D., 99

Fair Trade Commission, 88
Fox, Gerald, 8, 12, 15–17, 20–21

Henshaw, Paul, 69
Hokkaido University, meetings at, 24–30
Honda, Kotaro, 85
Horiuchi, Juro, 25–28, 30–34
Houston, W. V., 50

Ikeda, Hayato, 100
Ikeda, Kikunae, 85

Imperial Academy of Science, 26–27, 39, 47–48
Inouye, Jinkichi, 85
Itoda, Isao, 104

Japan Association of Science Liaison. *See* Science Liaison Group
Japan Development Bank, 70
Japanese Association for the Advancement of Science, 48
Japan Society for the Promotion of Science, 7, 26, 47–48
Jeffries, Zay, 56
Jewett, Frank, 10–11, 50–51

Kameyama, Naoto, 62, 64, 98–99
Kaneshige, Kankuro, 53–54, 62–64, 101–102
Katyama, Tetsu, 52
Kaya, Seiji, 31–33, 40, 83, 101

Lawrence, Ernest, 80

MacArthur, Douglas, 1, 51. *See also* Supreme Commander of Allied Powers
Manhattan Project, 3, 12
Ministry of Commerce and Industry, 37, 58–59, 76
Ministry of Education, 26, 37, 47–48, 55, 59, 75
Mukaibo, Takashi, 99–100, 102

Nagaoka, Hantaro, 66, 85
National Academy of Sciences, 44–45, 80–81
National Research Council of Japan, 22, 26–27, 39, 48
National Research Council (U.S.), 39–40, 44–45
National Research Foundation, 44
National Science Foundation, 96–98
Ni Project, 7

Nishibori, Eizaburo, 65–66
Nishina, Yoshio, 5–8, 62, 79–80, 82, 88–94
Nuclear Research Laboratory, 6

O'Brien, John, 8, 20–21, 52, 88
Office of Naval Research, 96
Okochi, Masatoshi, 85, 87, 89–90

Pacific Science Conference, seventh, 69
Patterson, Robert P., 9–10
Physical and Chemical Research Institute. *See* Riken
Physical Society of Japan, 17

Rabi, I. I., 56–57
Radiation Laboratory, MIT, 3, 15
Rebun-To expedition, 58
Reconstruction Finance Bank, 58
Reischauer, Edwin O., 100
Riken, 6, 79, 84–91
Robbins, W. J., 50
Royall, Kenneth, 11

Sagane, Ryokichi, 33–35, 40, 65–67, 70, 80
San Francisco Peace Treaty, 83, 92
Science Advisory Group, 56
Science Council of Japan, 54–62, 64–73 101–102
Science Liaison Group, 35–41, 45–47, 60, 62, 79
Science Organization Renewal Committee, 49–50, 52–55, 58, 60, 79
Science and Technology Agency, 58, 76–77
Science—the Endless Frontier (Bush), 44, 97
Scientific Advisory Group, 50–51, 81

Scientific Intelligence Survey of Japan, 7
Scientific and Technical Administration Commission, 54–55, 59, 61, 72–75, 82–83
Scientific and Technical Policy Comrades Association, 47
Security Treaty, U.S.-Japan, 101–102
Senshu, Kunio, 95–96
Sorenson, Royal, 50
Stakman, Elvin C., 56
Supreme Commander of Allied Powers, 46–47, 57, 59–60, 62, 64, 69–70, 88
Suzuki, Umetaro, 85

Tamiya, Horishi, 30–36, 38–40
Tomonaga, Shin'ichiro, 85

Ueda, Shunkich, 64

Wadachi, Kiyoo, 101
Wagatsuma, Sakae, 62
Waterman, Alan, 96–97

Yanibara, Tadao, 58
Yokoyama, Sumi, 92_93
Yoshida, Shigeru, 64
Yukawa, Hideki, 52, 85

Zaibatsu, 20, 87
Zaidan hojin, 89